797,885 Books
are available to read at

Forgotten Books

www.ForgottenBooks.com

Forgotten Books' App
Available for mobile, tablet & eReader

ISBN 978-0-266-10063-8
PIBN 10944082

This book is a reproduction of an important historical work. Forgotten Books uses state-of-the-art technology to digitally reconstruct the work, preserving the original format whilst repairing imperfections present in the aged copy. In rare cases, an imperfection in the original, such as a blemish or missing page, may be replicated in our edition. We do, however, repair the vast majority of imperfections successfully; any imperfections that remain are intentionally left to preserve the state of such historical works.

Forgotten Books is a registered trademark of FB &c Ltd.
Copyright © 2017 FB &c Ltd.
FB &c Ltd, Dalton House, 60 Windsor Avenue, London, SW19 2RR.
Company number 08720141. Registered in England and Wales.

For support please visit www.forgottenbooks.com

1 MONTH OF FREE READING

at
www.ForgottenBooks.com

By purchasing this book you are eligible for one month membership to ForgottenBooks.com, giving you unlimited access to our entire collection of over 700,000 titles via our web site and mobile apps.

To claim your free month visit: www.forgottenbooks.com/free944082

* Offer is valid for 45 days from date of purchase. Terms and conditions apply.

English
Français
Deutsche
Italiano
Español
Português

www.forgottenbooks.com

Mythology Photography **Fiction** Fishing Christianity **Art** Cooking Essays Buddhism Freemasonry Medicine **Biology** Music **Ancient Egypt** Evolution Carpentry Physics Dance Geology **Mathematics** Fitness Shakespeare **Folklore** Yoga Marketing **Confidence** Immortality Biographies Poetry **Psychology** Witchcraft Electronics Chemistry History **Law** Accounting **Philosophy** Anthropology Alchemy Drama Quantum Mechanics Atheism Sexual Health **Ancient History** **Entrepreneurship** Languages Sport Paleontology Needlework Islam **Metaphysics** Investment Archaeology Parenting Statistics Criminology **Motivational**

DAGGETT'S
School of Pharmacy Lecture Course.

A SHORT AND CONCISE SERIES OF

PRACTICAL LESSONS IN PHARMACY.

Designed for the use of Pharmacists, Physicians, Students intending
to take State Board of Pharmacy Examinations, and all others
who desire a practical knowledge of Drugs, Chemicals
and Pharmaceutical Compounds; together with
Tables of Incompatibilities, Latin Terms
used in Prescription Writing,
etc., etc., etc.

PREPARED BY

CHARLES H. DAGGETT, Ph. G.,

INSTRUCTOR OF

PHARMACY AND PHARMACEUTICAL CHEMISTRY.

Copyright 1898, by Charles H. Daggett.

SECOND EDITION, REVISED AND ENLARGED.

PROVIDENCE:
PRINTED BY JOHN F. GREENE COMPANY,
77 DYER STREET.
1899.

RS9
.D14

23125

TWO COPIES RECEIVED.

71580
Dec. 9, 98

PREFACE.

In offering DAGGETT'S SCHOOL OF PHARMACY LECTURE COURSE to the public, the author has been actuated by the complexity of text books in general, to produce a work which comprises in a concise and comprehensive form, the study of Pharmacy, Chemistry, and Materia Medica, and while the use of technical terms is unavoidable in a work of this kind, they are explained in such a manner as to render them easy to understand by the reader, after a little study. The writer does not claim originality in the branches taught, but only as to the arrangement of the work, which is the result of a number of years experience as a teacher.

It is invaluable to students wishing to pass State Board of Pharmacy examinations, on account of containing the greatest amount of information in the smallest amount of space; to the College Graduate, Registered Pharmacist, or Registered Assistant Pharmacist, as a reference; to Physicians in prescribing, containing, as it does, a simple explanation of the Metric System with equivalents, Latin terms used in prescription writing, incompatibles, doses, etc.

On account of the increased demand for copies of the first edition, which is exhausted, the second edition is offered to the public wholly on its merits.

THE AUTHOR.

DAGGETT'S
School of Pharmacy Lecture Course.

INTRODUCTION.

This course is not intended to be antagonistic to any method of study now pursued.

It is given for the benefit of those unable to attend a College of Pharmacy, who desire to obtain a practical education in the Theory and Practice of Pharmacy, and for those wishing to pass State Board of Pharmacy examinations.

Particular attention has been paid to making the explanations as plain and concise as possible, and a great amount of reading matter which usually accompanies a course of this kind is omitted.

For the College Graduate, Registered or Registered Assistant Pharmacist, it will be invaluable as a ready reference behind the prescription counter and in manufacturing, and for the physician in prescribing, containing, as it will, a simple explanation of the Metric System, Latin terms used in prescription writing, incompatibles, doses, etc.

The high standard of State Board of Pharmacy examinations makes it difficult for the average young man to prepare himself without the aid of a competent instructor.

For those taking the course, have regular hours for study, and let nothing, if possible, interfere with you.

Do not commit the lectures to memory, but endeavor to thoroughly understand them.

If possible, demonstrate the lectures by practical work and experiments in the store.

It is well if convenient, to quiz with a fellow student.

By faithful application to your studies it will take but a comparatively short time for you to master what may have at first seemed so difficult to understand.

In the arrangement of these lectures I am indebted to the following works: "U. S. Pharmacopœia," "U. S. Dispensatory," "Remington's Practice of Pharmacy," "Attfield's Chemistry," "Douglass and Prescott's Qualitative Analysis," "Avelling's Introduction to the Study of Botany."

C. H. DAGGETT, Ph. G.

LECTURE NO. 1.

PHARMACY is the art of preparing and dispensing drugs and medicines.

CHEMISTRY is the study of chemical force.

THERAPEUTICS is the application of medicine to diseases.

MATERIA MEDICA is that study which treats of the nature and properties of the substances used in medicine.

PHARMACOGNOSY is the study of the characters and identification of crude drugs.

MICROSCOPY is the study of the use and application of the microscope.

TOXICOLOGY is the study of poisons; their analysis and detection.

A PHAMACOPŒIA is a work of authority intended to regulate the strength and purity of medicinal substances and for the selection of remedies for the treatment of disease. The first Pharmacopœia was published by the Massachusetts Medical Society in 1808, the next one by the New York Hospital in 1816, and our present Pharmacopœia has been published every ten years since 1820.

A DISPENSATORY is a book, the object of which is to present an account of medicinal substances in the state in which they are brought into the shops, and to teach the modes in which they are prepared for use.

WEIGHT is gravity, ponderousness or downward pressure.

GRAVITY is the force that exists between all bodies.

WEIGHING is measuring the excess of terrestrial gravity of bodies in comparison with some body taken as a standard. The implement for weighing is called the balance.

The BALANCE is a perpendicular support, supporting a lever of the first class. The point of suspension is an acute angle of polished steel or agate, resting on a plane of hardened metal or agate. It must not only be a lever of the first class, but it must have the centre of gravity of the beam immediately below the point of suspension, and as near to it as possible.

Never touch the scale pans with the fingers.
Never overload the balance.
Always keep the balance in a glass case away from the dirt and dust.
Never handle the weights with the fingers.

Weights and Measures.

The Standard of avoirdupois weight is a brass weight weighing 7000 grains, divided into 16 ounces of 437.5 grains each.

The Standard of Troy or Apothecaries weight is a brass weight weighing 5760 grains, divided into 12 ounces of 480 grains each.

The Standard of wine measure is the gallon, containing 128 fluid ounces.

EQUIVALENTS.

One pound avoirdupois	equals	16 ounces.
One ounce avoirdupois	"	16 drachms.
One drachm avoirdupois	"	27.34 grains.
One pound troy		12 ounces.
One ounce troy		8 drachms.
One drachm troy		3 scuples.
One scruple		20 grains.
One gallon (Imperial)		8 pints.
One gallon (Wine)		8 pints.
One pint (Imperial)		20 fluid ounces.
One pint (Wine)		16 fluid ounces.
One fluid ounce (Imperial)	"	461 minims.
One fluid ounce (Wine)		480 minims.

RELATION OF WINE MEASURE TO TROY WEIGHT.

One minim	equals	0.95 grains.
One fluid drachm	"	56.96 grains.
One fluid ounce	"	455.70 grains.
One fluid ounce (Imperial)	"	437.50 grains.

RELATION OF TROY WEIGHT TO WINE MEASURE.

One grain	equals	1.05 minims.
One drachm	"	63.20 minims.
One ounce	"	505.60 minims.

APPROXIMATE EQUIVALENTS.

A teacup	equals	4 fluid ounces.
A wine glass	"	2 fluid ounces.
A tablespoon	"	½ fluid ounce.
A dessertspoon		2 fluid drachms.
A teaspoon		1 fluid drachm.

METRIC SYSTEM.

Is a French System based on multiples and subdivisions of ten.

The standard of surface measure is the METER, which is one forty-millionth part of a meridian of the earth through the poles.

The standard of solidity is the STER, which is a cube meter, or in other words, is the space occupied by a cube, each side of which is a meter in length.

The standard of capacity is the LITER, which is the cube of a decimeter or the one-tenth part of a meter. The Liter is too large for ordinary use, so a smaller subdivision is taken, which is the cube of the centimeter, or one-hundredth of a meter, and is called the CUBIC CENTIMETER.

The standard of weight is the GRAM, which is the weight of a cubic centimeter of water at 4°C.

Subdivisions and Multiples of the Metric System:

Milli	Meter	written	.001	pronounced as written.
Centi	"	"	.01	" Senti.
Deci	"		.1	" Dessy.
Meter			1.	Meeter.
Deka	"		10.	Decka.
Hecto	"		100.	" as written.
Kilo	"	"	1000.	" Killo.

The sub-divisions, deci, centi and milli, are taken from the Latin language, while the multiples, Deka, Hecto and Kilo are taken from the Greek language. These terms always mean the same, no matter what they are connected with.

Equivalents.

One Meter	equals		3 feet 3⅜ inches.
One Liter	"	1000.	cubic centimeters.
One cubic centimeter	"	16	minims.
One fluid ounce		30	cubic centimeters.
One gram		15.5	grains.
One troy ounce		31.1	grams.
One advoirdupois ounce		28.35	grams.
One advoirdupois pound		453.59	grams.
One troy pound	"	373.24	grams.

Specific Gravity.

Specific gravity is the comparative weight of the same volume of different bodies in comparison with water.

To take the specific gravity of liquids:

Use the specific gravity bottle, which is a bottle with a very narrow neck, fitted with a perforated stopper and adjusted usually, to hold 100 or 1000 grams. Divide the weight of the body, obtained by filling the bottle to the mark on the neck and weighing, by the weight of an equal bulk of water.

The HYDROMETER is also used to take the specific gravity of liquids. It consists of a graduated stem of glass with a bulb or bulbs at the bottom filled with mercury. The specific gravity of the liquid is ascertained by the depth to which it sinks in the liquid, the zero of the scale marking the depth to which it sinks in pure water.

To take the specific gravity of solids insoluble in, but heavier than water:

Weigh the substance in air: Weigh it in water by suspending it from the end of the beam of the balance by a thread and immersing in a flask of water, the flask not touching the scale pan. The loss in weight in water is the weight of an equal bulk of water. Divide the weight of the substance in air by the loss of weight in water and the result is the specific gravity of the solid.

To take the specific gravity of substances soluble in, but heavier than water: It must be weighed in some substance in which it is insoluble, as turpentine, oil, etc. Weigh it in the air, weigh it in turpentine, the specific gravity of which has been previously ascertained. The loss in weight in turpentine is the weight of an equal bulk of turpentine, then by comparison ascertain the weight of an equal bulk of water. Make the following proportion: The specific gravity of the turpentine is to the specific gravity of water as the weight of turpentine (represented by the loss in weight in turpentine) is to the weight of an equal bulk of water. Then divide the weight of the substance by the weight of the equal bulk of water, and the result is the specific gravity of the substance.

To take the specific gravity of bodies insoluble in, but lighter than water: It must be attached to something heavier than water, as brass, iron, etc. Weigh the substance in the air, attach to the piece of brass which has been previously weighed in water, and the loss noted. After attaching the brass to the substance weigh them both in water and subtract the loss due to the brass from the loss of both, which gives the weight of an equal bulk of water to the substance. Divide the weight in air by this weight and the result is the specific gravity of the substance.

THERMOMETERS:

A Thermometer is an instrument of glass with a graduated stem, having at the lower end a bulb filled with mercury, and is used to tell the amount of sensible heat that a body contains.

There are three kinds in use, namely: Fahrenheit, having the boiling point at 212° and freezing 32°; Centigrade, which is our official thermometer, having the boiling point at 100° and the freezing 0°, Reumer, having the boiling point at 80° and freezing 0°.

PHARMACEUTICAL TERMS.

DESSICATION is the process of depriving solid substances of moisture, and should be affected at as low a temperature as possible. Its object is to reduce the bulk of substances and to aid in comminuting and preserving them.

COMMINUTION is the process of reducing drugs to particles, or breaking up their state of aggregation.

LEVIGATION is the process of reducing substances to a state of minute division by triturating them after they have been made into a paste with water or some other liquid. The object of this process is to reduce such substances as zinc oxide or mercuric oxide to a fine state of subdivision, which cannot be performed by dry trituration in a mortar.

ELUTRIATION is the process of obtaining substances in fine powder by suspending an insoluble substance in water, allowing the heavier particles to fall to the bottom of the vessel and decanting off the liquid containing the lighter particles into another vessel and collecting them. Ex. Prepared Chalk.

TROCHISCATION is the process of making the pasty mass or magna obtained by elutriation into dry, conical masses.

PULVERIZATION BY INTERVENTION is the process of reducing substances to powder through the use of a foreign substance from which it is subsequently freed by some simple means. Ex. Camphor is powdered by rubbing with a little alcohol and the alcohol removed by evaporation.

BOILING is occasioned by the bubbles of vapor forming on the surface of a heated liquid. When the temperature of the liquid has reached the boiling point, 100° C. (212° F.) the bubbles pass off as steam.

VAPOR is a term applied to such substances as are condensed by pressure or cooled to a liquid state, this being their normal condition at ordinary temperature and pressure.

EVAPORATION is the conversion of a liquid or solid body into gas or vapor.

DISTILLATION is the employment of evaporation with the intention of condensing the vapor. This process is employed for the separation of volatile liquids from those less volatile, and is also employed for the purpose of purifying substances.

SIMPLE DISTILLATION is where a liquid is simply converted into vapor and the vapor condensed for the purpose of purification. Ex. Distillation of Water.

FRACTIONAL DISTILLATION is where different substances are obtained from the same body, only at different temperatures. Ex. Carbolic acid.

DESTRUCTIVE DISTILLATION is where solid organic substances are submitted to such a high degree of heat that their structure is changed and new bodies are formed. Ex. Acetic Acid.

SUBLIMATION is the conversion of volatile solids into vapor and condensing the vapor. Ex. Sulphur, benzoic acid, etc.

FUSION is the conversion of solid bodies into a liquid state without changing their characteristics. Ex. Lunar caustic.

CALCINATION is the process of driving off all the volatile portions of a body. Ex. Calcined Magnesia.

GRANULATION is the process of heating the solution of a chemical substance with constant stirring until the moisture has evaporated and a coarse grained powder results.

EXSICCATION is the process of depriving a solid crystalline substance of its water of crystallization or moisture by heating it strongly. The object is to increase the strength and thereby fit it for special application. Ex. Dried Sulphate of Iron.

IGNITION is the process of converting a body to ashes by the action of strong heat, one of its uses being to convert organic bodies into such a condition that the amount of organic matter that they contain may be shown.

DEFLAGRATION is the process of heating one inorganic substance with another capable of yielding oxygen (usually nitrate or a chlorate) causing decomposition with violent, noisy or sudden combustion.

CARBONIZATION is the process of subjecting organic substances to a strong heat without the access of air. Ex. Animal Charcoal.

SOLUTION is a mechanical combination of the molecules of a body producing a clear liquid. SIMPLE SOLUTION is one in which the dissolved body has undergone no sensible change of properties, and in which the mechanical forces are so balanced that a clear liquid results. COMPOUND SOLUTION is one in which more than one body is acted on by the same solvent. COMPLEX SOLUTION is where an entire change of properties takes place and we cannot recover the bodies by any mechanical means.

CRYSTALLIZATION is the process whereby substances are caused to assume a regular geometric form.

A SATURATED SOLUTION is one in which the solvent has dissolved all of the body that it can.

EFFLORESCENCE is the property a body has of losing its water of crystallization by exposure to the air.

DELIQUESCENCE is the property a body has of absorbing moisture from the air.

DECANTATION is the process of pouring off from above, the supernatant liquid.

DIALYSIS is the passage of crystalloids through a porous septum, which is usually parchment paper. It is the separation of crystalloids from colloids.

The DIALYSATE is the liquid in the dialyzer.

The DIFFUSATE is the liquid that has passed through the dialyzer.

COLLOIDS are bodies which pass slowly through a porous septum.

CRYSTALLOIDS are bodies which pass rapidly through a porous septum.

FILTRATION is the separation of insoluble matter from liquids by passing through a porous septum, such as gray and white filter paper, etc.

STRAINING or COLATION is the same as filtration, only passing through a coarser septum, as fine muslin, cotton flannel, etc.

DECOLORIZATION is the process of taking color from a body. The best material for the purpose is animal charcoal.

CLARIFICATION is the process of separating from liquids without the use of strainers or filters, solid substances that interfere with their transparency.

A SEDIMENT is solid matter separated from the liquid in which it is suspended by the action of gravity.

A PRECIPITATE is solid matter separated from a solution by heat, light or chemical action. The separation of the precipitate is called PRECIPITATION, the liquid left above is called the SUPERNATANT LIQUID, and whatever is added to cause a precipitate to form is called the PRECIPITANT.

A MAGNA is an amorphous precipitate, holding water mechanically, which results from straining or filtering. Ex. Ferric Hydrate.

MACERATION is a preliminary process accompanying percolation, and sometimes alone. It consists in allowing the solvent to come in contact with the cells of a drug for a specified length of time, usually from twenty-four to forty-eight hours.

DIGESTION is practically the same as maceration, only at a higher temperature.

EXPRESSION is the process of separating solids from liquids by force.

PERCOLATION is the process of passing menstruum equally through a drug contained in a percolator.

PERCOLATORS are vessels of different sizes and shapes, and made of different substances, as tinned iron, glass, stoneware, etc., intended to hold the drug, while the menstruum is allowed to pass through it, extracting its active principles. The percolator most suitable for the quantities as given in the U. S. P., should be nearly cylindrical, or slightly conical, with a funnel-shaped termination at the smaller end.

REPERCOLATION consists in passing the undersaturated, or weaker percolates from one portion of the drug through another, and again passing the undersaturated, or weaker percolate from this second portion through a third, and so on.

FERMENTATION is a process of changing bodies to their elements from which new bodies are formed.

A FERMENT is a body capable of acting on other bodies at a certain temperature, producing fermentation.

The MENSTRUUM is a solvent or liquid which passes through the drug, extracting its active constituents.

An EXTRACT is a body obtained by evaporating solutions of vegetable principles, and is of a solid, or semi-solid consistence. Ex. Extract of Gentian.

LECTURE NO. 2.

An INSPISSATED JUICE is an extract obtained by evaporating the expressed juice of a vegetable substance to a solid, or semi-solid consistence. Ex. Aloes.

A GUM is a vegetable substance usually an exudation, insoluble in alcohol; either soluble in water, forming a mucilaginous mass, or swelling up in contact with it. The gums consist of three proximate principles namely; ARABIN, found mostly in acacia, and is soluble. BASSORIN, found mostly in tragacanth, and is insoluble. CERASIN, found in cherry gum, and is insoluble.

A SUGAR is an organic substance having a sweet taste, crystallizable, soluble in water and dilute alcohol, but insoluble in alcohol and ether. Ex. Cane Sugar.

A RESIN is a solid or semi-solid exudation from a vegetable substance, insoluble in water and soluble in alcohol, ether and chloroform. Ex. Mastic.

A GUM RESIN is a natural mixture of gum and resin and forms an emulsion when triturated with water. Ex. Ammoniac.

An OLEORESIN is a mixture of volatile oil and resin. Ex. Copaiba.

A BALSAM is an oleoresin, containing, besides oil and resin, either benzoic or cinnamic acids, or both. Ex. Balsam of Tolu.

A VOLATILE OIL is a limpid liquid substance usually obtained by distillation from plants. It has a strong odor, resembling that of the plant from which it is obtained. The solid crystalline substance deposited by volatile oils on standing, is called a STEAROPTEN, and from some plants is called a CAMPHOR, from its resemblance to true camphor. The liquid remaining after the solid portion is removed is called ELEOPTEN. The volatile oils are divided into three classes: Non-Oxygenated Oils, consisting of carbon and hydrogen. Ex. Turpentine. Oxygenated Oils, containing carbon, hydrogen and oxygen. Ex. Oil of Cinnamon. Sulphuretted Oils, containing sulphur. Ex Mustard.

A FIXED OIL is a liquid or semi-liquid obtained by expression from vegetable and animal substances. The fixed oils contain a liquid principle called Olein, and two concrete substances called Palmitin and Stearin. A Volatile Oil is distinguished from a fixed oil by not leaving a greasy stain on paper.

An ALKALOID is a vegetable base and always contains nitrogen, or it is one of a group of organic bodies containing nitrogen. Chemically alkaloids are either Amines, containing carbon, nitrogen and hydrogen, or Amides, containing carbon, nitrogen, hydrogen and oxygen. They are mostly soluble in alcohol, chloroform, benzin, etc., insoluble in water, and unite with acids to form salts.

The Latin names of alkaloids end in "ina," while the English names end in "ine."

As a general treatment in case of alkaloidal poisoning use emetics and the stomach pump.

A GLUCOSIDE is a body which, under the action of a ferment splits up into glucose and some other body peculiar to itself.

The Latin names of glucosides end in "inum," while the English names end in "in."

Therapeutical Terms.

An ABSORBENT is a drug used to produce absorption of diseased tissue. Ex. Iodine.

An ALTERATIVE is a medicine used to modify nutrition so as to overcome morbid processes. Ex. Arsenic.

An ANÆSTHETIC is a medicine used to produce insensibility. Ex. Ether.

An ANALGESIC is a medicine used to relieve pain. Ex. Opium.

An ANAPHRODISIAC is a medicine that lowers sexual functions.

An ANODYNE is a medicine used to relieve pain externally. Ex. Soap Liniment.

An ANTACID is a medicine used to neutralize acid in the stomach and intestines. Ex. Magnesia.

An ANTHELMINTIC is a medicine used to expel worms from the stomach or intestines. Ex. Male Fern.

An ANTILITIC is a medicine used to relieve stone in the bladder. Ex. Lithium Carbonate.

An ANTIPYRETIC is a medicine used to reduce bodily temperature in fevers. Ex. Phenacetine.

An ANTISEPTIC is a substance which prevents putrefaction. Ex. Iodoform.

An ANTISPASMODIC is a medicine used to relieve nervous irritability and minor spasms.

An APERIENT is a mild purgative. Ex. Seidlitz Powder.

An APHRODISIAC is a medicine used to stimulate sexual functions. Ex. Phosphorus.

An AROMATIC is a medicine that has a fragrant taste or odor and stimulates the gastro intestinal mucous membrane. Ex. Cardamom.

An ASTRINGENT is a medicine that contracts tissue. Ex. Nutgall.

A CARDIAC STIMULANT is a medicine that stimulates the heart's action. Ex. Digitalis.

A CARDIAC SEDATIVE is a medicine that reduces the heart's action. Ex. Dilute Hydrocyanic Acid.

A CARMINATIVE is a medicine used to relieve flatulence or colic. Ex. Spirit of Anise.

A CATHARTIC is a medicine that causes the contents of the stomach to pass off through the bowels.

A CHALYBEATE is a medicine in which iron predominates. Ex. Basham's Mixture.

A CHOLAGOGUE is a medicine which causes a flow of bile. Ex. Chrysarobin.

A DEMULCENT is a medicine that produces a bland sensation and soothes irritated surfaces. Ex. Acacia.

A DEPRESSO MOTOR is a medicine that depresses muscular activity. Ex. Conium.

A DETERGENT is a medicine used to cleanse ulcers, wounds, etc. Ex. Zinc Sulphate.

A DIAPHORETIC or SUDORIFIC is a medicine that increases the secretions of the skin. Ex. Aconite.

A DISINFECTANT is a substance that has the power of destroying disease germs. Ex. Carbolic Acid.

A DIURETIC is a medicine that increases the secretions of the kidneys. Ex. Buchu.

A DRASTIC is a powerful purgative. Ex. Calomel.

An EMETIC is a medicine that causes vomiting. Ex. Mustard.

An EMMENAGOGUE is a medicine that stimulates menstruation. Ex. Savine.

An EMOLLIENT is a substance used to mechanically soften and protect tissues. Ex. Lard.

An EPISPASTIC is a medicine causing a blister. Ex. Mustard.

An ERRHINE is a medicine that increases the secretions of the nose without producing sneezing, as a rule. Ex. Vapor of Ammonia.

An EXCITO MOTOR is a medicine that stimulates muscular activity. Ex. Nux Vomica.

An EXPECTORANT is a medicine that acts on the pulmonic mucous membrane and increases or alters its secretions. Ex. Guaiac.

A HÆMOSTATIC is a medicine that arrests hemorrhages. Ex. Monsell's Solution.

A HYPNOTIC is a medicine that causes sleep without previous cerebral excitement. Ex. Chloral.

A MIDRIATIC is a medicine used to dilate the pupil of the eye. Ex. Belladonna.

A MYOTIC is a medicine used to contract the pupil of the eye. Ex. Physostigmine.

A NARCOTIC is a medicine that produces sleep by exciting the brain. Ex. Opium.

An OXYTOXIC is a medicine that contracts the uterus. Ex. Ergot.

A PURGATIVE is a medicine that acts as a powerful cathartic. Ex. Chrysarobin.

A SEDATIVE is a medicine that depresses vital force. Ex. Dilute Hydrocyanic Acid.

A SIALAGOGUE is a medicine that increases the flow of saliva. Ex. Mezereum.

A STIMULANT is a medicine that increases functional activity. Ex. Alcohol.

A STOMACHIC is a medicine that stimulates the stomach. Ex. Gentian.

A STYPTIC is the same as hæmostatic.

A TONIC is a medicine that permanently increases the tone of the system by stimulating nutrition. Ex. Gentian.

Chemical Terms.

An ELEMENT is a substance which cannot by any known means be resolved into a simpler form of matter. Elements are either metallic or non-metallic and are capable of being separated from their compounds by the action of an electric current and collect around the negative and positive poles. Ex. If chloride of zinc be decomposed by an electric current the zinc will collect at the negative pole and the chlorine at the positive pole. The elements that collect at the negative pole are called ELECTRO POSITIVE, and are metallic elements. Those that collect around the positive pole are called ELECTRO NEGATIVE, and are mostly non-metallic elements.

Prof. W. Ripley Nichols gives a list of electro positive and electro negative elements as follows :

Electro Positive: Gold, Platinum, Silver, Mercury, Copper, Tin, Lead, Cobalt, Nickel, Iron, Zinc, Manganese, Aluminum, Magnesium, Calcium, Sodium, Potassium.

Electro Negative: Oxygen, Sulphur, Nitrogen, Fluorine, Chlorine, Bromine, Iodine, Phosphorus, Arsenic, Boron, Carbon, Antimony, Silicon, Hydrogen.

A COMPOUND is a combination of two or more elements.

COHESION is that force by which particles of like bodies are held together. It is great in solids, small in liquids, and apparently absent in gases.

ADHESION is that force by which particles of unlike bodies are held together.

CHEMICAL AFFINITY is that power which binds the atoms of a compound so closely together that they seem to lose all individuality.

It produces an entire change of properties in the bodies over which it is exerted.

A CHEMICAL SYMBOL is a capital letter, or a capital and a small letter. 1. It is shorthand for the name of the element. 2. It represents an atom of the element. 3. It stands for a constant weight of the element. 4. It represents single and equal volumes of gaseous elements.

An ATOM is a particle so small that it cannot be subdivided, or it is the smallest particle of matter that can exist in a combined state. The weight of an atom is called ATOMIC WEIGHT, and is the proportion in which elements combine together by weight.

A MOLECULE is the smallest particle of matter that can exist in a free state, and is composed of atoms. The weight of a molecule is called MOLECULAR WEIGHT, and is the sum of the atomic weights entering into the molecule.

COMBUSTION is that variety of chemical combination, in which the chemical union is so intense as to produce light and heat.

A RADICAL is a group of atoms or a single atom which retains its form when transferred from one molecule to another, being a leading constituent of each. Ex. Hydrochloric acid. HCl. " H " is the base and " Cl " the radical.

An ACID is a salt of hydrogen, commonly possessing a sour taste, turning blue litmus red, and containing hydrogen capable of being replaced by a base forming a salt. Ex. Nitric acid. HNO_3.

An OXYACID is one whose radical contains oxygen. Ex. Nitric acid. HNO_3.

A HYDRACID is one whose radical contains no oxygen. Ex. Hydrochloric acid. HCl.

An ANHYDRIDE is an acid from which the elements of water have been removed, and its chemical properties altered. Ex. If we remove the elements of water (H_2O) from sulphuric acid (H_2SO_4) we have SO_3 the anhydride of sulphuric acid.

A BASE is a body capable of replacing the hydrogen of an acid to produce a salt. Ex. Metals.

A SALT is a body formed by the action of an acid on a base, or it is a body formed by substituting the whole or a part of the hydrogen in an acid by a metal. Ex. By replacing the hydrogen in sulphuric acid by its equivalent of iron forms sulphate of iron. $FeSO_4$.

A NORMAL SALT is formed by replacing all the hydrogen of the acid by its equivalent of metal. Ex. Sulphate of Iron.

An ACID SALT is formed by replacing a part of the hydrogen in the acid by its equivalent of metal. Ex. If we replace a part of the hydrogen in sulphuric acid by its equivalent of potassium we have acid sulphate of potassium ($KHSO_4$).

A BASIC SALT is formed by the substitution of a metal in part for the hydrogen of the acid, and in part for the half or the whole of the hydrogen in water, Bi, with $\frac{HNO_3}{H_2O}$ form Bi ONO$_3$ (Subnitrate of Bismuth), or, it may be formed by replacing a part of the acid radical by its equivalent of oxygen. Ex. Bi+3HNO$_3$=Bi(NO$_3$)$_3$ (normal bismuth nitrate). NO$_3$ is a monad radical, and if it is to be replaced by an atom of oxygen (which is a diad) two molecules of NO$_3$ will have to be removed to make room for the oxygen thus: Bi(NO$_3$)$_3$—(NO$_3$)$_2$=Bi NO$_3$+O=Bi ONO$_3$. Ex.

QUANTIVALENCE represents the value of atoms in relation to one of hydrogen. Each element has a certain number of bonds equal to a certain number of atoms of hydrogen. The number of bonds an element has is marked thus: K′, Ca″, As‴, showing that potassium replaces one atom of hydrogen in an acid, calcium two atoms, and arsenic three atoms.

The value of all the elements may be found in a table in the last part of the U. S. D., 17th Edition.

Elements having one bond are termed MONADS and called UNIVALENT. Those having two bonds, DIADS and called BIVALENT. Those having three bonds, TRIADS and called TRIVALENT.

ISOMERIC BODIES are bodies having the same chemical composition, yet differing in properties. Ex. Starch and Dextrine.

ALLOTROPIC BODIES are bodies similar in composition and constitution, yet differing in properties. Ex. Phosphorus and Red Phosphorus.

POLYMERIC BODIES are those having the same percentage composition, yet differing in molecular weight and properties. They are multiples. Ex. Aldehyde C$_2$H$_4$O and Paraldehyde C$_6$H$_{12}$O$_3$.

AMORPHOUS BODIES are those without chrystalline form.

ISOMORPHOUS BODIES are those that crystallize in the same form.

DIMORPHOUS BODIES are those that crystallize in two distinct forms, belonging to different systems.

ALCOHOLS are bodies derived from hydrocarbons by replacing one or more atoms of hydrogen by hydroxyl (HO). They are called monatomic, diatomic or triatomic alchohols, according as one, two or three atoms of hydrogen are replaced. Ex. CH$_4$ methane, CH$_3$OH methyl alcohol: C$_2$H$_6$ ethane, C$_2$H$_5$OH ethyl alcohol: etc.

The hydrocarbons from which alcohols are obtained are formed from marsh gas or methane (CH$_4$) by the successive addition of CH$_2$, the series being called a HOMOLOGOUS SERIES. Ex. CH$_4$ methane, C$_2$H$_6$ ethane, C$_3$H$_8$ propane, C$_4$H$_{10}$ butane, C$_5$H$_{12}$ amane.

A HETEROLOGOUS SERIES is where the different series of bodies as alcohols, aldehydes, acids, ethers, etc., are given together.

A PRIMARY ALCOHOL is one in which the carbon atom in combination with hydroxyl is also directly connected with one other carbon atom. Ex. Butyl Alcohol, Primary C_4H_9OH, or $CH_3(CH_2)_2CH_2OH$.

$$H-\underset{\underset{H}{|}}{\overset{\overset{H}{|}}{C}}-\underset{\underset{H}{|}}{\overset{\overset{H}{|}}{C}}-\underset{\underset{H}{|}}{\overset{\overset{H}{|}}{C}}-\underset{\underset{H}{|}}{\overset{\overset{H}{|}}{C}}-OH$$

It may be seen that the characteristic primary alcohol group is CH_2OH.

A SECONDARY ALCOHOL is one in which the carbon atom in combination with hydroxyl is also directly connected with two other carbon atoms. Ex. Butyl Alcohol, Secondary C_4H_9OH, or $CH_3CH_2CH_3CHOH$.

$$CH_3-\underset{\underset{H}{|}}{\overset{\overset{H}{|}}{C}}-\underset{\underset{H}{|}}{\overset{\overset{CH_3}{|}}{C}}-OH$$

It may be seen that the characteristic group of the secondary alcohol is CHOH.

A TERTIARY ALCOHOL is one in which the carbon atom in combination with hydroxyl is also connected with three other carbon atoms. Ex. Butyl Alcohol, Tertiary. C_4H_9OH, or $(CH_3)_3COH$.

$$CH_3-\underset{\underset{CH_3}{|}}{\overset{\overset{CH_3}{|}}{C}}-OH$$

AROMATIC ALCOHOLS are bodies formed by replacing an atom of the side group of one of the benzene series by its equivalent of hydroxyl. Ex. $C_6H_5CH_3$. Toluene.

$C_6H_5CH_2OH$. Benzyl Alcohol.

PHENOLS are bodies formed from the aromatic series of hydrocarbons called the benzene series by replacing a hydrogen atom of the nucleus of benzene by its equivalent of hydroxyl. Ex. C_6H_6. Benzene.
C_6H_5OH Phenol.

ALDEHYDES are bodies obtained by removing one or more molecules of hydrogen from a primary alcohol, whereby the alcohol group CH_2OH is changed to COH, which is the characteristic aldehyde group. Ex. CH_3CH_2OH. Ethyl alcohol.

CH_3COH. Ethyl Aldehyde.

ORGANIC ACIDS are bodies formed by oxidizing aldehydes, whereby the aldehyde group is changed to COOH, which is the characteristic acid group. They are also formed by replacing one or more

molecules of hydrogen in an alcohol by its equivalent of oxygen. Ex. CH_3CH_2OH. Ethyl Alcohol.
CH_3COOH. Acetic Acid.

ESTERS or COMPOUND ETHERS are bodies formed from acids by replacing the basic hydrogen for alcohol radicals.
Ex. $HC_7H_5O_3$. Salicylic Acid.
$CH_3C_7H_5O_3$. Methyl Salicylate.

KETONES are bodies formed by the oxidation of secondary alcohols, which contain the group CHOH as before stated, whereby one molecule of hydrogen is taken away and we have the characteristic ketone group CO remaining. They may also be obtained by substituting an alcohol radical for the hydrogen in the aldehyde group COH.
Ex. CH_3COH. Ethyl Aldehyde.
CH_3COCH_3. Acetone.

AMIDES are bodies obtained from ammonia by replacing the hydrogen by acid radicals. They are called monamides, diamides, or triamides, according as one, two or three atoms of hydrogen are replaced.
Ex. NH_3 Ammonia.
$NH_2C_2H_3O$. Acetamide.

AMINES are bodies obtained from ammonia by replacing the hydrogen by alcohol radicals. Ex. NH_3 Ammonia.
$NH_2C_2H_5$ Ethylamine.

ALKALAMIDES are bodies obtained by replacing the hydrogen in ammonia, partly by acid radicals and partly by alcohol radicals.
Ex. NH_3. Ammonia.
$NHC_2H_5C_2H_3O$. Ethylacetamide.

HYDROCARBONS are bodies containing carbon and hydrogen. Ex. C_6H_6 Benzene.

CARBOHYDRATES are bodies containing carbon, hydrogen and oxygen, the hydrogen and oxygen being in the proportion in which they unite to form water. Ex. $C_6H_{10}O_5$ Starch.

A RATIONAL FORMULA is one which represents the different grouping of the atoms in the molecule. Ex. CH_3CH_2OH. Ethyl Alcohol.

An EMPIRICAL FORMULA is one which simply represents the number of atoms of each element in the molecule, or it may be defined as the simplest expression of the composition of a substance. Ex. C_2H_6O. Ethyl Alcohol.

GLYCERIDES are compound ethers of the triatomic alcohol glycerin $C_3H_5(OH)_3$, and of the several fatty acids, oleic, palmitic, stearic, etc. Ex. $C_3H_5(C_{17}H_{33}O_2)_3$. Olein or Oleate of Glyceryl.

FATS and FIXED OILS are bodies containing the above glycerides, Oleate, Palmitate and Stearate of Glyceryl, therefore glyceryl bears the same relation to fats and fixed oils that ethyl does to alcohol.

LECTURE NO. 3.

WAXES differ from fats and fixed oils in being compound ethers of the higher monatomic alcohol, as cetyl alcohol, etc.

SOAPS are mixed oleates, palmitates and stearates of various bases.

PREFIXES: The following definitions are those as given by Prof. Patch, in a paper read before the Massachusetts Pharmaceutical Association in 1892.

MONO: From monos, single or alone, commonly indicates that one atom of the element or residue it qualifies enters into the compound. Ex. PbO. Monoxide of Lead.

PROTO: From protos, first, usually means the lowest compound in a series. Ex. FeO. Protoxide of Iron.

BI or BIN: From bis, twice, indicates twice as much in value or saturating power when used to indicate the proportion of acid in a compound. Ex. $KHCO_3$. Bicarbonate of Potassium. It may also indicate the degree of of oxidation as $HgCl_2$. Bichloride of Mercury.

DEUTO: From duo, two, or deuteros, second; is also used to indicate the presence of two atoms of a negative element. Ex. CrO_2. Chromium Deutoxide, Binoxide or Dioxide.

DI: From dis, twice, is not only used to indicate the proportion of negative elements, but also the proportion of base. Ex. The normal quinine sulphate is known as the Disulphate, while the acid sulphate is known as the Bisulphate.

TRI: From Greek tria, three, and TER, from Latin ter, thrice, both indicate three atoms of a negative element or their equivalent. Ex. Fe_2O_3. Teroxide or trioxide of Iron.

SESQUI: From Latin sequi, one and a half, indicates the ratio of one and a half of oxygen as negative element to one of positive. Ex. Fe_2O_3. Sesquioxide of Iron.

TETRA: From Greek tetratos, fourth, is usually employed to indicate the proportion of negative element. Ex. K_2O_4. Tetraoxide of Potassium.

PENTA: From Greek pente, five.

HEPTA: Seven.

HYPER: Means excess, higher, above, etc. It is usually contracted to PER. Perchlorides, Peroxides, Persulphates, etc., indicate combinations representing higher oxidation. Ex. Fe_2O_3. Peroxide of Iron.

SUPER: Means excess, and is usually applied to excess of acid in relation to base. Ex. $KHC_4H_4O_6$. Supertartrate or bitartrate of potassium.

HYPO: Signifies under, beneath, lower. It is used to indicate a lower degree of oxidation than normal. Ex. N_2O. Nitrous oxide combines with water (H_2O) to form HNO. Hyponitrous Acid.

SUB: Under, or below, qualifies the acid radical, as in subsulphate of iron. ($Fe_4O(SO_4)_5$) we have a deficiency of one molecule of sulphuric acid as compared with normal ferric sulphate. It is also an OXYSULPHATE and a BASIC SULPHATE.

ORTHO: Straight; that is regular or ordinary in combination, orthophosphate, the commonly made salt. Na_2HPO_4.

META: Beyond, between, reverse, change, etc. Ex. Metaphosphoric Acid is Phosphoric Acid H_3PO_4 less H_2O or HPO_3.

PYRO: Fire; that is, produced by heat. $HC_7H_5O_5$ gallic acid, less CO_2 becomes $C_6H_6O_3$. Pyrogallic acid.

PARA: Alongside of, or equal to. Ex. C_2H_4O aldehyde and three molecules is $C_6H_{12}O_3$ paraldehyde.

OUS: Is a suffix used to denote lower oxides and their compounds, or lower oxyacids. Ex. H_2SO_3 Sulphurous Acid.

IC: Is appended to higher oxides and their compounds, or applied to many acids formed by direct union of the positive or basic, and one negative or acid element. HCl. Hydrochloric Acid.

SALT ENDINGS: A salt whose name ends in (ate) is made from an acid whose name ends in (ic). A salt whose name ends in (ite) is made from an acid whose name ends in (ous). A salt whose name ends in (ide) is as a rule made by the direct combination of an element with a base.

A LIQUID is a substance, the molecules of which move about each other so freely that it readily assumes and retains the form of any vessel in which it is placed.

A LIMPID LIQUID is one in which the molecules move very freely. Ex. Water.

A VISCID LIQUID is one in which the molecules move very sluggishly. Ex. Syrup.

A SOLID is a substance the molecules of which are more or less immovable, though not probably in absolute contact. Ex. Wood.

A GAS is a substance the molecules of which are so far apart that they seem to lose all attraction, and to have the property of repulsion to such an extent that they are only prevented from receding to a still greater extent by the pressure of the surrounding matter. Ex. Hydrogen.

ANHYDROUS BODIES are compounds from which water has been taken away.

WATER OF CRYSTALLIZATION is the water that a body contains in its natural state.

DIFFUSION is the power bodies have of naturally mixing with each other. This power of diffusion is particularly characteristic of gases.

A MECHANICAL MIXTURE is one in which the dissolved or mixed bodies have undergone no chemical change. Ex. Citrated Caffeine.

A CHEMICAL COMPOUND is one in which the dissolved bodies have undergone a chemical change and new bodies are formed. Ex. Syrup Iodide of Iron.

BOTANICAL TERMS.

A LEAF is the expanded part of the stem.

A BULB is an underground stem with fleshy scales. Ex. Squill.

A ROOT is the descending axis of a plant. Ex. Gentian.

A FRUIT is the matured ovary. Ex. Anise.

A RHIZOME is a creeping stem or branch growing beneath the surface of the soil, or partly covered by it. Ex. Male Fern.

The BARK is the covering of the stem outside the wood. Ex. Cinnamon.

The RIND is the outside covering of the fruit. Ex. Orange.

A FLORET is a little flower on one of the flowers of a cluster. Ex. Marigold.

A CORM is an underground stem, with central axis and only one layer of scales. Ex. Colchicum.

A STAMEN is one of the male reproductive organs of the flower.

A STIGMA is the top of the pistil, which is the female reproductive organ of a flower. Ex. Saffron.

A CAPSULE is a pod.

A STROBILE is a multiple fruit in the form of a cone or head. Ex. Hop.

A TUBER is a short, thick, globular underground stem or root stalk, with much starch. Ex. Jalap, Potato, etc.

An ARILLODE is a false coat or fleshy growth, covering the seed or fruit. Ex. Mace.

A LEAFLET is one of the divisions or blades of a compound leaf. Ex. Senna.

The CALYX is the outer set of the floral envelope of the flower.

The COROLLA is the inner set of the floral envelope of the flower and is located just inside the calyx.

The PETALS are the leaves or divisions of the corolla. Ex. Red Rose.

A STYLE is a part of the pistil which bears the stigma. Ex. Corn Silk.

Definitions of the classes of preparations of the U. S. P. 1890, the preparations themselves being treated on later, under their respective heads.

ACETA (Medicinal Vinegars) are solutions of medicinal substances in dilute acetic acid. They were formerly made with vinegar as the solvent, but on account of the variable strength and the presence of extractive matters in the vinegar, the preparations were liable to decomposition and dilute acetic acid of definite strength (6%) was substituted. There are two official, namely: Opium and Squill, each containing 10% of the drug.

AQUÆ (Medicated Waters) are preparations consisting of water holding volatile or gaseous substances in solution. They are prepared by direct solution, as in chloroform water, intermediate solution as in peppermint water, and by distillation, as in rose water.

Intermediate solution consists in saturating the water with the volatile substance by the aid of some other substance. The U. S. P., 1880, used cotton for this purpose, but it was found to be objectionable on account of not thoroughly saturating the water and also in allowing particles of oil to pass through into the water. The U. S. P., 1890, has adopted precipitated phosphate of calcium as the best medium, in that it is insoluble and contains no soluble impurities. There are seventeen waters official, namely: Ammonia, stronger ammonia, bitter almond, anise, orange flower, camphor, chlorine, chloroform, cinnamon, creosote, distilled, fennel, oxygenized, peppermint, spearmint, rose, stronger rose.

CERATA (Cerates) are unctuous substances consisting of oil or lard, mixed with wax, spermaceti, or resin, to which various medicaments are frequently added. Their consistence is intermediate between that of ointments and plasters. They should be kept in a cool place in well covered jars. It has been shown that cerates prepared with yellow wax keep unaltered much longer than those made with white wax. There are six official, namely: Simple, camphor, cantharides, spermaceti, lead subacetate, resin.

CONFECTIONES (Confections) are preparations having the form of a soft solid, in which one or more medicinal substances are mixed with saccharine matter with a view to their preservation or more convenient administration. Their consistence should not be so soft, on the one hand, as to allow the ingredients to separate, nor so firm, on the other, as to prevent them from being swallowed without mastication. There are two official, namely: Rose, senna.

DECOCTA (Decoctions) are solutions of vegetable principles obtained by boiling the substances containing them with water. Their strength in the U. S. P., 1880, was 10%, while in the U. S. P., 1890, it has been changed to 5%.

Decoctions when prescribed and the strength not specified should be made 5%. There are two official, namely: Cetraria and sarsaparilla compound.

ELIXIRIA (Elixirs) are aromatic, sweetened, spirituous preparations containing small quantities of active medicinal substances. There are two official, namely: Aromatic, phosphorus.

EMPLASTRA (Plasters) are solid compounds intended for external application, adhesive at the temperature of the human body, and of such a consistence as to render the aid of heat necessary in spreading them. They are prepared for use by spreading upon leather, linen, or muslin, according to the purpose for which they are intended. There are thirteen official, namely: Ammoniac with mercury, arnica, belladonna, capsicum, iron, mercurial, isinglass, opium, burgundy pitch, cantharidal pitch, lead, resin and soap.

EMULSA (Emulsions) are liquid preparations in which oleaginous or resinous substances are suspended in watery fluids by the intervention of gum, yolk of eggs or other viscid matter. Natural Emulsions are those found in nature ready formed, as in the milky juices of plants, animal milk, yolk of egg, etc. Manufactured Emulsions are those made artificially by various methods. Emulsions are made by one of two methods, namely:

English Method, which consists- in placing in a mortar the emulsifying agent, mucilage, yolk of egg, etc., and adding the oil and water alternately in small quantities, constantly stirring until emulsification is complete.

Continental Method, which consists in placing four parts of oil in a mortar, adding one part of powdered gum, stirring, and when a uniform mixture is made, adding two parts of water all at once and emulsifying, then adding more water, if necessary, according to the strength required. There are four official, namely: Ammoniac, almond, asafœtida and chloroform.

EXTRACTA (Extracts) are solid preparations, resulting from the evaporation of solutions of vegetable principles, obtained either by exposing a dried drug to the action of a solvent, or by expressing the juice from the fresh plant. The strength of these preparations bears no definite relation to the drug, the amount of extract obtained depending on the mode of preparation, the solvent employed and the nature of the drug. The more aqueous the menstruum the more extract obtained, and the more alcoholic the less extract obtained.

They should be kept in well-closed vessels. There are thirty-three official, namely: Aconite, aloes, arnica root, belladonna alcoholic, Indian hemp, cimicifuga, cinchona, colchicum root, colocynth, colocynth compound, conium, digitalis, ergot, euonymus, gentian, licorice, licorice pure, logwood, henbane, iris, jalap, juglans, krameria, leptandra, nux vomica, opium, physostigma, podophyllum, quassia, rhubarb, stramonium seed, taraxacum, and uva ursi.

EXTRACTA FLUIDA (Fluid Extracts) are practically concentrated tinctures, in which one cubic centimeter or fluid gram of the finished fluid represents the strength of a gram of the drug.

The U. S. P. process for making fluid extracts is as follows: One thousand grams of the drug reduced to the proper fineness are moistened, packed, macerated and percolated to exhaustion. The first portion, from 700 to 900 C. c. of the percolate is reserved, the weaker percolate is evaporated at a low temperature to the consistence of a soft extract, dissolved in the reserved portion, and enough menstruum added to make the finished preparation measure 1000 C. c. Fluid Extracts should be kept away from the light and in a room in which there is not much variation in temperature. There are eighty-nine official, namely: Aconite, Canadian hemp, arnica root, aromatic, asclepias, aspidosperma, bitter orange, belladonna root, buchu, calamus, columbo, cannabis indica, capsicum, castanea, chimaphila, chirata, cimicifuga, cinchona, coca, colchicum root, colchicum seed, conium, convallaria, cubeb, cusso, cypripedium, digitalis, dulcamara, ergot, eriodictyon, eucalyptus, euonymus, eupatorium, frangula, gelsemium, gentian, geranium, glycyrrhiza, cotton root, grindelia, gurana, hamamelis, hydrastis, hyoscyamus, ipecac, iris, krameria, lappa, leptandra, lobelia, lupulin, matico, menispermum, mezereum, nux vomica, pareira, poke root, pilocarpus, podophyllum, wild cherry, quassia, rhamus purshiana, rhubarb, rhus glabra, rose, rubus, rumex, sabina, sanguinaria, sarsaparilla, sarsaparilla compound, squill, scoparius, skullcap, senega, senna, serpentaria, spigelia, stillingia, stramonium seed, taraxacum, triticum, uva ursi, valerian, veratrum viride, viburnum opulus, viburnum prunifolium, xanthoxylum, ginger.

The process of repercolation is much to be preferred in making fluid extracts in that the use of heat is done away with and the liability of injuring the active principle not occurring. The only objection to this process is that of carrying numerous reserves in stock.

GLYCERITA (Glycerites) are solutions of medicinal substances in glycerin. They are sometimes called glyceroles. There are six official, namely: Carbolic acid, tannic acid, starch, boroglycerin, hydrastis, yolk of egg.

INFUSA (Infusions) are aqueous solutions obtained by treating with water, without the aid of boiling, vegetable products only partially soluble in it. The water used in making them may be hot or cold, as the case requires, cold water being used when the active principle is highly volatile, or when injured by heat. The strength of the U. S. P. 1880 infusions was 10%.

Infusions when prescribed and the strength not specified, should be made of 5% strength as follows: The substance is placed in a suitable vessel provided with a cover, boiling water poured upon it and macerated for half an hour, covered tightly. It is then strained and enough water added to make the required quantity. As infusions do not keep well they should be made in small quantities as required. There are four official, namely: Cinchona, digitalis, senna compound, wild cherry.

LINIMENTA (Liniments) are preparations intended for external use, of such consistence as to render them conveniently applicable to the skin by gentle friction with the hand. There are nine official, namely: Ammonia, belladonna, lime, camphor, chloroform, soap, soft soap, mustard compound, turpentine.

LIQUORES (Solutions) are aqueous solutions without sugar in which the substance acted on is wholly soluble in water, excluding those in which the dissolved matter is gaseous or very volatile, as in the aquæ. There are twenty-four official, namely: Arsenous acid, ammonium acetate, arsenic and mercuric iodide, lime, ferric acetate, ferric chloride, ferric citrate, iron and ammonium acetate, ferric nitrate, ferric subsulphate, ferric sulphate, mercuric nitrate, iodine compound, magnesium citrate, lead subacetate, lead subacetate diluted, potassa, potassium arsenite, potassium citrate, soda, soda chlorinated, sodium arsenate, sodium silicate and zinc chloride.

MISTURÆ (Mixtures) are preparations in which insoluble subtances, whether solid or fluid, are suspended in watery fluids by the intervention of gums, yolk of egg, or other viscid matter. There are four official, namely: Chalk, iron compound, licorice compound, rhubarb and soda.

MUCILAGINES (Mucilages) are aqueous solutions of gum or of substances closely allied to it. There are four official, namely: Acacia, sassafras, tragacanth and elm.

OLEATA (Oleates) are solutions of certain bases in oleic acid, and are made by triturating the solid substance with the oleic acid until it is dissolved. There are three official, namely: Mercury, veratrine and zinc.

OLEORESINÆ (Oleoresins) are preparations that consist of principles, which when extracted by means of ether, retain a liquid or semi-liquid state upon the evaporation of the menstruum, and at the same time have the property of self-preservation, differing in this respect from the fluid extracts, which require the presence of alcohol in order to prevent decomposition. There are six official, namely: Aspidium, capsicum, cubeb, lupulin, pepper and ginger.

PILULÆ (Pills) are globular masses of a convenient size for swallowing. Deliquescent substances should not be made into pills, and those which are efflorescent should be previously deprived of their water of crystallization. There are fifteen official, namely: Aloes, aloes and asafetida, aloes and iron, aloes and mastic, aloes and myrrh, antimony compound, asafetida, cathartic compound, vegetable cathartic, ferrous carbonate, ferrous iodide, opium, phosphorus, rhubarb and rhubarb compound.

PULVERES (Powders). The form of powder is convenient for the exhibition of substances which are not given in very large doses, and are not very disagreeabe to the taste, have no corrosive property and do not deliquesce rapidly. There are nine official, namely: Antimonial, aromatic, chalk compound, effervescing compound, licorice compound, ipecac and opium, jalap compound, morphine compound, rhubarb compound.

SPIRITUS (Spirits) are alcoholic solutions of volatile principles formerly in general procured by distillation, but now prepared frequently by simply dissolving the volatile principle in alcohol or dilute alcohol. There are twenty-five official, namely: Ether, ether compound, nitrous ether, ammonia, ammonia aromatic, bitter almond, anise, orange, orange compound, camphor, chloroform, cinnamon, frumenti, gaultheria, glonoin, juniper, juniper compound, lavender, lemon, peppermint, spearmint, myrcia, nutmeg, phosphorus and vini gallici.

SUPPOSITORIA (Suppositories) are solid bodies intended to be introduced into the rectum with a view of either evacuating the bowels by irritating the mucous membrane of the rectum, or of producing a specific effect on the neighboring parts, or on the system at large.

The official directions for making suppositories are as follows: Having weighed out the medicinal ingredient or ingredients and the quantity of oil of theobroma required, according to the kind of suppository to be prepared, mix the medicinal portion (previously brought to the proper consistence, if necessary) with a small quantity of oil of theobroma, by rubbing them together, and add the mixture to the remainder of the oil of theobroma, previously melted and cooled to $35°$ C ($95°$ F). Then mix thoroughly without applying more heat, and immediately pour the mixture into suitable moulds. The moulds must be kept cold by being placed on ice, or by immersion in ice water before the melted mass is poured in.

Unless otherwise specified suppositories should have the following weights and shapes corresponding to their uses. Rectal Suppositories should be cone shaped, and of a weight of about one gram ($15\frac{1}{2}$ grains). Urethal Suppositories should be pencil shaped, and of a weight of about one gram ($15\frac{1}{2}$ grains). Vaginal Suppositories should be globular and of a weight of about three grams (46 grains). There is one official, namely: Glycerin.

SYRUPI (Syrups) are concentrated solutions of sugar in aqueous fluids, either with or without medicinal impregnation. There are thirty-two official, namely: Simple, acacia, citric acid, hydriodic acid, garlic, althæa, almond, orange, orange flowers, calcium lactophosphate, lime, ferrous iodide, iron quinine and strychnine phosphates, hypophosphites, hypophosphites with iron, ipecac, krameria, lactucarium, tar, wild cherry, rhubarb, rhubarb aromatic, rose, rubus, raspberry, sarsaparilla compound, squill, squill compound, senega, senna, tolu, ginger.

LECTURE NO. 4.

TINCTURÆ (Tinctures) are alcoholic solutions of medicinal substances, prepared by maceration, digestion or percolation. The official tinctures are made by percolation. There are seventy-one official, namely: Aconite, aloes, aloes and myrrh, arnica flowers, arnica root, asafetida, bitter orange peel, sweet orange peel, belladonna leaves, benzoin, benzoin compound, bryonia, calendula, columbo, cannabis indica, cantharides, capsicum, cardamom, cardamom compound, catechu compound, chirata, cimicifuga, cinchona, cinchona compound, cinnamom, colchicum seed, crocus, cubeb, digitalis, ferric chloride, nutgall, gelsemium, gentian compound, guaiac, guaiac ammoniated, hops, hydrastis, henbane, iodine, ipecac and opium, kino, krameria, lactucarium, lavender compound, lobelia, matico, musk, myrrh, nux vomica, opium, opium camphorated, opium deodorized, physostigma, pyrethrum, quassia, quillaja, rhubarb, rhubarb aromatic, rhubarb sweet, sanguinaria, squill, serpentaria, stramonium seed, strophanthus, sumbul, tolu, valerian, valerian ammoniated, vanilla, veratrum viride, ginger.

TROCHISCI (Troches) sometimes called LOZENGES, are small, dry, solid masses, usually of a flattened shape, consisting for the most part of powders incorporated with sugar and mucilage. There are fifteen official, namely: Tannic acid, ammonium chloride, catechu, chalk, cubeb, iron, licorice and opium, ipecac, krameria, peppermint, morphine and ipecac, potassium chlorate, santonin, sodium bicarbonate, and ginger.

UNGUENTA (Ointments) are fatty substances, softer than cerates, of a consistency like that of butter, and such that they may be readily applied to the skin by inunction. There are twenty-three official, namely: Simple, carbolic acid, tannic acid, rose water, belladonna, chrysarobin, diachylon, nutgall, mercurial, mercury ammoniated, mercuric nitrate, yellow mercuric oxide, red mercuric oxide, iodine, iodoform, tar, lead carbonate, lead iodide, potassium iodide, stramonium, sulphur, veratrine, zinc oxide.

VINA MEDICATA (Medicated Wines) are solutions of medicinal substances in white wine. There are eight official, namely: Antimony, colchicum root, colchicum seed, ergot, iron bitter, ferric citrate, ipecac, opium.

INORGANIC PHARMACY.

ACIDS.

ACIDUM ARSENOSUM (Arsenous acid) As_2O_3. Arsenic Trioxide, White Arsenic, Arsenious Acid, Arsenious Anhydride.

It is not a true acid, but an anhydride, obtained by roasting arsenical ores and purified by sublimation.

There are two varieties in the market, namely: Glassy and opaque. The glassy is the more soluble and on exposure to the air for some time it becomes opaque. The powdered acid as found in the shops is often adulterated with calcium sulphate, chalk, or powdered lime, which may be detected by heating the mixture sufficiently to vaporize the arsenous acid, and the impurities will remain behind. It is official in solution of arsenous acid and solution arsenite of potassium.

It is an alterative. Dose: $\frac{1}{30}$ to $\frac{1}{20}$ of a gr. (0.002 to 0.003 gm.)

The antidote to poisoning by it is hydrated oxide of iron with magnesia.

The most common method of identifying arsenic is MARSH'S TEST, which consists in generating hydrogen in a flask with sulphuric acid, zinc and water; the gas is passed through a glass tube drawn out to a small point; after the gas has been coming over for a short time add some of the suspected solution containing arsenic to the flask and ignite the gas at the small point on the tube and allow the flame to play on a porcelain plate; if arsenic or antimony is present a black metallic mirror will be deposited on the plate. The two metals may be distinguished by adding to the spot a drop or two of fuming nitric acid and heating: arsenic will thus be converted into soluble arsenic acid, precipitated brick red by nitrate of silver; antimony, on the other hand, into insoluble antimonic acid. Another way to distinguish one from the other is to add a little solution of chlorinated lime, which will dissolve the arsenic without affecting the antimony.

LIQUOR ACIDI ARSENOSI (Solution of Arsenous Acid) Hydrochlorate Solution of Arsenic.

Contains 1% of arsenous acid and 5% of hydrochloric acid to aid in its solubility in water. Used for the same purpose as Fowler's Solution.

Dose: 2 to 8 minims (0.12 to 0.5 C. c.).

LIQUOR POTASSII ABSENITIS (Solution of Potassium Arsenite) Fowler's Solution.

Contains 1% of arsenous acid with potassium bicarbonate, water and compound tincture of lavender, which is added to impart a color so that it may not be mistaken for water.

It is incompatible with the salts of iron, magnesium and calcium.

Used as an alterative. Dose: 5 drops (0.3 C. c.).

ACIDUM BORICUM (Boric Acid) H_3BO_3. Obtained from the lakes in Tuscany which contain it, by evaporating and crystallizing. It is also obtained from borax by the action of hydrochloric acid. Soluble in 25.6 parts of water and 15 of alcohol, and the addition of hydrochloric acid increases its solubility in water.

It is antiseptic. Dose: 10 grains (0.647 gm.). Official in Glycerite of Boroglycerin.

GLYCERITUM BOROGLYCERINI (Glycerite of Boroglycerin). Glycerite of Glyceryl Borate, Solution of Boroglyceride.

Prepared by heating 310 grams of boric acid with 460 grams of glycerin. When dissolved continue the heat at 150°C (302°F) frequently stirring and breaking the film which gathers on the surface. When the mixture has been reduced to the weight of 500 grams add 500 grams of glycerin.

ACIDUM CHROMICUM (Chromic Acid) Chromic Anhydride. CrO_3

It is not a true acid, but an anhydride, obtained by acting on potassium bichromate with sulphuric acid.

It should be kept in glass-stoppered bottles and care used not to bring it in contact with organic matter as glycerin, alcohol, sugar, tannic acid, cork, etc., as violent and dangerous decomposition is apt to occur.

It is antiseptic and disinfectant. Dose: ¼ grain (0.016 gm.).

ACIDUM HYDROBROMICUM DILUTUM (Diluted Hydrobromic Acid) HBr.

Contains 10% by weight of absolute acid and 90% by weight of water.

Prepared by acting on potassium bromide with tartaric acid, also by passing sulphuretted hydrogen through a mixture of bromine and water (the sulphur of the sulphuretted hydrogen separating). It unites with water and alcohol in all proportions. Its action is the same as that of potassium bromide.

Dose: 1 to 2 fluid drachms (3.75 to 7.50 C. c.).

ACIDUM HYDROCHLORICUM (Hydrochloric Acid) HCl. Spirit of Sea Salt, Marine Acid, Muriatic Acid, Chlorhydic Acid.

Contains 31.9% of absolute acid and 68.1% water, and should be kept in dark amber glass-stoppered bottles.

Prepared by the action of sulphuric acid on sodium chloride and passing the gas into water. Also obtained in the process of converting sodium chloride into sodium carbonate.

The acid and its preparations are incompatible with alkalies, metallic oxides, sulphides, potassium tartrate, tartar emetic, silver nitrate, lead compounds, etc.

When exposed to the air the escaping acid gas unites with the moisture in the air, causing white fumes.

It is tonic, refrigerant and antiseptic. Dose: 5 to 10 minims (0.3 to 0.6 C. c.) Antidotes: Dilute alkaline solutions, magnesia, soap, etc. It is official in dilute hydrochloric acid, nitrohydrochloric acid and dilute nitrohydrochloric acid.

ACIDUM HYDROCHLORICUM DILUTUM (Diluted Hydrochloric Acid).

Contains 10% by weight of absolute acid, and is prepared by mixing together 100 grams of hydrochloric acid and 219 grams of water.

Dose: 15 to 30 minims (0.92 to 1.85 C. c.).

ACIDUM HYDROCYANICUM DILUTUM (Diluted Hydrocyanic Acid) HCN. Prussic Acid, Cyanhydric Acid.

Contains 2% by weight of absolute acid and 98% of water. It is prepared by heating together potassium ferrocyanide and diluted sulphuric acid and passing the gas into water. Also may be prepared by acting on silver cyanide with hydrochloric acid.

It is very poisonous and unstable, and should be kept in well corked bottles in a dark place. Dose: 1 to 3 minims (.06 to 0.18 C.c.) Used as a sedative.

Its antidotes are nitrite of amyl and ammonia vapors.

ACIDUM NITROHYDROCHLORICUM (Nitrohydrochloric Acid). Aqua Regia, so named on account of its property of dissolving gold.

Prepared by mixing together 180 C.c. of nitric acid and 820 C.c. of hydrochloric acid in a capacious vessel, and, when effervescence has ceased pouring the product into dark amber-colored bottles with glass stoppers, which should not be more than half filled and kept in a cool place.

When these acids are mixed together, nitrosyl chloride, free chlorine and water are formed. It has an orange color, soon changing to a golden yellow and having a strong odor of chlorine. It should be kept in a cool dark place on account of its liability of losing chlorine by heat, and to have its chlorine converted into hydrochloric acid by the action of light and the decomposition of water, which would be an objection to its use.

The strong acid is preferable to the dilute. It should never be compounded with strong alcoholic liquids, as explosion might result from the formation of gases.

It is used in hepatic affections and in dyspepsia. Dose: 3 to 6 drops (0.18 to 0.36 C.c.) The strong acid is preferable to the dilute.

ACIDUM NITROHYDROCHLORICUM DILUTUM (Dilute Nitrohydrochloric Acid). Contains 4% of nitric acid, 18% of hydrochloric acid with water. The acids should be mixed in a capacious vessel, and when effervescence has ceased, add the distilled water. The product should be kept in dark amber-colored, glass-stoppered bottles, in a cool place.

Between dilute nitric and hydrochloric acids no reaction takes place therefore the strong acids should be mixed first.

Dose: 10 to 20 drops (0.6 to 1.25 C.c.).

ACIDUM HYPOPHOSPHOROSUM DILUTUM (Diluted Hypophosphorus Acid). A liquid composed of about 10% by weight of absolute acid (HPH_2O_2) and about 90% of water.

Prepared by boiling phosphorus with milk of lime, and decomposing the calcium hypophosphite with a strong acid. Also prepared by the action of tartaric acid dissolved in dilute alcohol on potassium hypophosphite dissolved in water. The alcohol aids in the precipitation of the potassium bitartrate.

It is of great value in preserving preparations containing iodides liable to decomposition through exposure to light and air. It mixes in all proportions with water or alcohol. Care should be taken not to triturate the salts of this acid with organic substances, since explosion is liable to occur.

Used in preparing syrup of hypophosphites. Dose: 10 to 30 minims (0.62 to 1.85 C.c.).

ACIDUM NITRICUM (Nitric Acid) HNO_3. Spirit of Nitre. Aqua Fortis.

Contains 68% by weight of absolute acid and 32% of water, and should be kept in dark amber glass-stoppered bottles. Prepared by distilling together a mixture of potassium nitrate or sodium nitrate and sulphuric acid.

It is an extremely sour and corrosive liquid. It acts on organic tissues, such as nails, skin, etc., turning them yellow.

Great care should be used in transporting it, for if the strong acid comes in contact with a quantity of vegetable substances like hay, tow, excelsior, paper, etc. fire is apt to occur.

It is a tonic, antiseptic and astringent. Dose: 5 to 10 minims (0.3 to 0.6 C.c.).

Antidotes: Alkaline solutions, soap, magnesia, chalk in large doses, olive and almond oil. Official in the dilute acid.

ACIDUM NITRICUM DILUTUM (Diluted Nitric Acid).

Contains 10% by weight of absolute acid, and is prepared by mixing 100 grams of nitric acid with 580 grams of distilled water.

Its medical properties are the same as that of the strong acid. Dose: 20 to 40 drops or minims (1.25 to 2.5 C.c.).

ACIDUM PHOSPHORICUM (Phosphoric Acid) H_3PO_4. Orthophosphoric Acid.

Contains 85% by weight of absolute acid and 15% of water, and should be kept in glass-stoppered bottles.

Prepared by acting on phosphorus with nitric acid in the presence of water, as follows:

Place the nitric acid diluted with water in a flask, add the phosphorus to the flask, connect it with a vertical glass condenser, boil so that all the condensed products shall return to the flask, continue boiling until the phosphorus has disappeared, remove the condenser, concentrate the liquid until orange colored vapors are no longer given off, and mix with distilled water to make the required strength.

If orthophosphoric acid be heated for a considerable time to 215°C (419°F) two molecules of the acid lose one molecule of water and yield pyrophosphoric acid thus: $2H_3PO_4$ or $H_6P_2O_8 - H_2O = H_4P_2O_7$. At a red heat the orthophosphoric acid is converted into metamosphoric acid thus: $H_3PO_4 - H_2O = HPO_3$, which is a transparent ice-like solid.

Metaphosphoric acid coagulates albumen and gives white, gelatinous, uncrystallizable precipitates with salts of lime, barium and silver: Pyrophosphoric acid does not coagulate albumen, and, though it causes a white precipitate with silver nitrate, must first be neutralized: Orthophosphoric acid does not coagulate albumen, but after neutralization, with silver nitrate (with which it then precipitates) throws down a yellow precipitate of silver phosphate. Metaphosphoric acid should be used in prescriptions with pyrophosphate of iron as the orthophosphoric acid is incompatible with it. Orthophosphoric acid should be used in prescriptions containing chloride of iron, as metaphosphoric acid is incompatible with it.

Sulphuric acid may be detected in it by diluting the acid with water, adding barium chloride test solution, which causes a precipitate. Hydrochloric acid may be detected by adding to the diluted acid, silver nitrate test solution, which causes a precipitate. Pyrophosphoric and Metaphosphoric acids may be detected by the addition of an equal volume of tincture of ferric chloride, causing a precipitate on standing.

Used in making the dilute acid.

ACIDUM PHOSPHORICUM DILUTUM (Diluted Phosphoric Acid).

Prepared by mixing together 100 grams of phosphoric acid and 750 grams of distilled water, and contains 10% by weight of absolute acid. On long keeping it becomes stringy, owing to the formation of a microscopic plant.

It is tonic and refrigerant. Dose: 20 to 60 drops (1.25 to 3.75 C.c).

ACIDUM SULPHURICUM (Sulphuric Acid) H_2SO_4. Oil of Vitriol, Vitriolic Acid. Contains 92.5% of absolute acid and 7.5% of water, and should be kept in glass-stoppered bottles.

Prepared by burning sulphur or iron pyrites (sulphide of iron) and allowing the product of combustion SO_2, to mix with nitrous fumes obtained by the decomposition of nitrate of potash, which changes the SO_2 into SO_3 and this uniting with steam, forms H_2SO_4.

It unites with water in all proportions and much heat is evolved when the two fluids are mixed together. The most common impurities are arsenous acid and lead sulphate.

It acts as a powerful caustic on living tissues. It is never used full strength internally. Antidotes: Large quantities of magnesia or solution of soap. Official in the dilute acid and aromatic sulphuric acid.

ACIDUM SULPHURICUM DILUTUM (Diluted Sulphuric Acid).

Prepared by mixing together 100 grams of sulphuric acid and 825 grams of distilled water, and contains 10% by weight of absolute acid.

It is tonic, refrigerant and astringent. Dose: 10 to 30 drops (0.6 to 1.9 C.c.).

ACIDUM SULPHURICUM AROMATICUM (Aromatic Sulphuric Acid). Elixir Vitriol.

Contains 20% of the official acid, with oil of cinnamon, tincture of ginger and alcohol, and should be kept in glass-stoppered bottles.

It is a tonic and astringent. Dose: 10 to 30 drops (0.6 to 1.9 C.c.)

ACIDUM SULPHUROSUM (Sulphurous Acid) H_2SO_3.

Contains about 6.4% by weight of absolute acid gas, (SO_2) and about 93 6% of water.

Prepared by heating sulphuric acid and charcoal and passing the gas into water. When exposed to the air it slowly absorbs oxygen and is converted into sulphuric acid.

It is antiseptic and germacide. Dose: 1 fluid drachm (3.75 C.c.) largely diluted with water.

Aluminum Compounds.

ALUMEN (Alum) $Al_2K_2(SO_4)_4$, $24 H_2O$, Potassa Alum, Aluminum and Potassium Sulphate.

Obtained principally from alum stone, which is a native mixture of aluminum sulphate and potassium sulphate. Also obtained from aluminous shist, which is a mixture of iron disulphide with aluminia, silica and bituminous matter. On exposure to the air the crystals are liable to absorb ammonia, and acquire a whitish coating. Soluble in water and glycerin, but insoluble in alcohol.

It is incompatible with the alkalies and their carbonates, lime and lime water, magnesia and its carbonate, potassium tartrate, and lead acetate.

It is a powerful astringent. Dose: 10 to 60 grains (0.65 to 3.9 grams). Used in making dried alum.

ALUMEN EXSICCATUM (Dried Alum) Alumen Ustum, Burnt Alum. $Al_2K_2(SO_4)_4$.

Prepared by heating alum on a water bath until it loses its water of crystallization. It attracts moisture on exposure to the air. Soluble in water. It is astringent and mild escharotic.

ALUMINI HYDRAS (Aluminum Hydrate) $Al_2(OH)_6$. Hydrated Alumina.

Prepared by precipitating a solution of alum with one of sodium carbonate. Insoluble in water and alcohol, but soluble in hydrochloric or sulphuric acids.

It is feebly astringent and dessicant.

ALUMINI SULPHAS (Aluminum Sulphate). $Al_2(SO_4)_3$, 16 H_2O.

Prepared by acting on aluminum hydrate with sulphuric acid. Soluble in water and insoluble in alcohol. It is antiseptic and mild caustic.

Ammonium Compounds.

AQUA AMMONIÆ (Ammonia Water). Solution of Ammonia. An aqueous solution of ammonia, containing not less than 10% by weight of the gas (NH_3).

Prepared by heating ammonium chloride with lime and passing the gas into water. Also made by diluting the stronger water. Incompatible with acids, and with acidulous and many earthy and metallic salts.

It is stimulant, irritant, and caustic. Used in making ammonia liniment and aromatic spirit of ammonia.

LINIMENTUM AMMONIÆ (Ammonia Liniment). Volatile Liniment. Contains ammonia water, cotton seed oil, and a little alcohol to prevent immediate separation. The ammonia reacts with the oil to form a soap, which is partly dissolved and partly suspended in water, producing a white, opaque emulsion. Used as a rubifacient.

SPIRITUS AMMONIÆ AROMATICUS (Aromatic Spirit of Ammonia). Compound Spirit of Ammonia.

Contains ammonium carbonate, ammonia water, oil of lemon, oil of lavender flowers, oil of nutmeg, alcohol and water. The carbonate of ammonia is a mixture of carbamate and bicarbonate of ammonia. The ammonia water converts the bicarbonate into carbamate, which is soluble in alcohol, wihle the bicarbonate is not.

It gradually becomes darker on standing any length of time.

It is a stimulant antacid. Dose: 30 to 60 drops (1.9 to 3.75 C.c.).

AQUA AMMONIÆ FORTIOR (Stronger Ammonia Water). Contains 28% by weight of the gas in water.

Prepared on a large scale from gas liquor, which is principally ammonium sulphide, by distilling and converting into ammonium sulphate, with sulphuric acid, then distilling the sulphate with milk of lime.

SPIRITUS AMMONIÆ (Spirit of Ammonia). Ammoniated Alcohol. Contains 10% by weight of the gas.

Prepared by heating stronger water of ammonia and passing the gas into alcohol.

It is stimulant and anti-spasmodic. Dose: 10 to 30 drops (0.6 to 1.9 C.c.).

AMMONII BENZOAS (Ammonium Benzoate) $NH_4C_7H_5O_2$.

Prepared by dissolving benzoic acid in ammonia water mixed with water, evaporating, keeping the ammonia in slight excess, and setting aside to crystallize. It gradually loses ammonia on exposure to the air. Soluble in water and alcohol.

It is a stimulant diuretic. Dose: 10 to 30 grains (0.65 to 1.95 gms.).

LECTURE NO. 5.

AMMONII BROMIDUM (Ammonium Bromide) NH_4Br.

Prepared by dissolving bromine in ammonia water, evaporating and granulating. Also by double decomposition between solutions of ammonium sulphate and potassium bromide.

On exposure to the air it gradually becomes yellowish, on account of partial decomposition by which hydrobromic acid is liberated. Soluble in water and alcohol. Incompatible with acids, acid salts and spirit of nitrous ether.

Used for the same purposes as potassium bromide. Dose: 1 to 2 drachms a day (3.9 to 7.8 gms.).

AMMONII CARBONAS (Ammonium Carbonate) NH_4HCO_3, NH_4, NH_2CO_2. Called Sal Volatile, Volatile Salt.

Prepared by subliming together a mixture of ammonium chloride or sulphate with calcium carbonate, and should be kept in well stoppered bottles, in a cool place.

When exposed to the air for any length of time it loses ammonia and carbonic acid gas, and becomes covered with a whitish coating of ammonium bicarbonate (acid carbonate). It is soluble in water.

It is cardiac stimulant. Dose: 5 grains (0.33 gms.).

Official in aromatic spirit of ammonia and solution ammonium acetate.

LIQUOR AMMONII ACETATIS (Solution of Ammonium Acetate) Spirit of Mindererus.

Prepared by adding ammonium carbonate to dilute acetic acid until the solution is slightly acid in reaction. It should be prepared fresh when wanted for use, because its acid becomes decomposed and a portion of ammonium carbonate is generated.

It is diaphoretic. Dose: $\frac{1}{2}$ to $1\frac{1}{2}$ fluid ounce (15 to 45 C.c.). Official in solution iron and ammonium acetate.

LIQUOR FERRI ET AMMONII ACETATIS (Solution of Iron and Ammonium Acetate) Basham's Mixture.

Contains tincture of ferric chloride, dilute acetic acid, solution of ammonium acetate, aromatic elixir, glycerin and water. The glycerin is used as a preservative.

It is chalybeate and astringent. Dose: $\frac{1}{2}$ to 1 fluid oz. (15 to 30 C.c.).

AMMONII CHLORIDUM (Ammonium Chloride) NH_4Cl, Muriate of Ammonia, Sal Ammoniac, Hydrochlorate of Ammonia.

Prepared by saturating gas liquor with sulphuric acid, evaporating, and subliming the ammonium sulphate obtained with sodium chloride.

Also obtained from bone spirit, a by-product in the destructive distillation of bones, in the manufacture of bone-black; it may also be obtained from stale urine, coal soot, guano, peat and bituminous schist.

Permanent in the air. Soluble in water and insoluble in alcohol.

It has the stimulant properties of ammonia. Dose: 5 to 10 grains (0.33 to 0.65 gms.) Official in the troche.

TROCHISCI AMMONII CHLORIDI (Troches of Ammonium Chloride).

Contains ammonium chloride, extract of licorice, tragacanth, sugar and syrup of tolu. Each troche contains 1½ grs. of ammonium chloride.

AMMONII IODIDUM (Ammonium Iodide) NH_4I. Prepared by dissolving a mixture of ammonium sulphate and potassium iodide in boiling water, cooling and adding alcohol to aid in precipitation of the potassium sulphate. It should be kept in small, well-stoppered bottles, protected from light.

The salt becomes yellow or yellowish brown on exposure to air and light, owing to the loss of ammonia and the elimination of iodine. Soluble in water and alcohol.

It is resolvent. Dose: 3 to 5 grains (0.2 to 0.33 gms.)

AMMONII NITRAS (Ammonium Nitrate) NH_4NO_3, Inflammable Nitre. Prepared by saturating nitric acid with ammonium carbonate, filtering and evaporating.

Deliquescent when exposed to the air. Soluble in water and alcohol.

Used in preparing nitrous oxide gas or Laughing Gas.

AMMONII VALERIANAS (Ammonium Valerianate) $NH_4C_5H_9O_2$.

Prepared by saturating valerianic acid with gaseous ammonia, and allowing to crystallize. It should be kept in well-stoppered bottles.

Deliquescent in moist air and efflorescent in dry air. Soluble in water and alcohol.

Used in nervous diseases. Dose: 2 to 8 grains (0.13 to 0.52 gms.).

ANTIMONY COMPOUNDS.

ANTIMONIUM (Antimony) Sb. It is a silver white metallic element obtained principally in Germany and France and exists in nature as a sulphide Sb_2S_3, oxysulphide $Sb_2O_3Sb_2S_3$, antimony teroxide Sb_2O_3, and in a free state. It forms three compounds with oxygen, antimony trioxide Sb_2O_3, antimony tetroxide Sb_2O_4, antimony pentoxide Sb_2O_5.

ANTIMONII ET POTASSII TARTRAS (Antimony and Potassium Tartrate) $2KSbOC_4H_4O_6,H_2O$, Tartar Emetic, Tartarized Antimony.

Prepared by boiling together oxide of antimony and potassium bitartrate with water, evaporating and crystallizing.

Soluble in water and insoluble in alcohol. It is incompatible with acids, alkalies and their carbonates, some of the earths and metals, calcium chloride, lead acetate, lead subacetate, astringent infusions and decoctions, but these astringent bodies, except galls, do not render it inert, though they lessen its activity. In case of poisoning wash out the stomach with solutions of tannic acid, then give stimulants, opiates, etc.

It often contains as impurities cream of tartar, calcium tartrate, iron, sulphates and chlorides.

Dose, as an alterative: $\frac{1}{32}$ to $\frac{1}{15}$ of a grain (0.002 to 0.004 gms.), as a diaphoretic or expectorant: $\frac{1}{12}$ to $\frac{1}{6}$ of a grain (0.005 to 0.01 gms.), as a sudorific: $\frac{1}{6}$ to $\frac{1}{4}$ of a grain, (0.01 to 0.016 gms.), as an emetic: $\frac{1}{2}$ grain (0.03 gm.).

Official in compound syrup of squill and wine of antimony.

VINUM ANTIMONII (Wine of Antimony) Antimonial Wine.

Prepared by dissolving tartar emetic in boiling distilled water, and adding to white wine containing a little alcohol.

It contains .4% tartar emetic. Dose: 10 to 30 drops (0.6 to 1.9 C.c.).

ANTIMONII OXIDUM (Antimony Oxide) Sb_2O_3, Antimony Teroxide, Antimony Trioxide.

Prepared by adding sodium carbonate to a solution chloride of antimony. Permanent in the air. Almost insoluble in water and insoluble in alcohol. Soluble in hydrochloric, and tartaric acid, or in a boiling solution of potassium bitartrate.

It is alterative Dose: 3 grains (0.2 gms.) Official in antimonial powder.

PULVIS ANTIMONIALIS (Antimonial Powder) James's Powder, Compound Powder of Antimony.

Contains 33% of antimony oxide and 67% of precipitated calcium phosphate.

As often found it is insoluble in water, but usually a small portion, consisting of antimonite and acid calcium phosphate dissolves in boiling distilled water.

It is alterative, diaphoretic and purgative. Dose: as a diaphoretic, 3 to 8 grains (0.2 to 0.52 gms.).

ANTIMONII SULPHIDUM (Antimony Sulphide) Sb_2S_3. Antimony Trisulphide, Black Antimony.

Obtained from the native sulphide, called "antimony ore" by fusion.

Much of the black antimony of commerce contains no antimony whatever, but consists of powdered charcoal and marble mixed together. Permanent in the air. Insoluble in water and alcohol, but soluble in hydrochlorine acid.

It is diaphoretic and alterative. Dose: 10 to 30 grains (0.65 to 1.95 gms.).

ANTIMONII SULPHIDUM PURIFICATUM (Purified Antimony Sulphide). Sb_2S_3.

Prepared by freeing antimony sulphide from the coarser particles by elutriation, macerating with ammonia water to remove any arsenic or copper that may be present, washing and drying.

Permanent in the air. Insoluble in water or alcohol. Used in preparing Kermes's Mineral.

ANTIMONIUM SULPHURATUM (Sulphurated Antimony) Sb_2S_3, Kermes's Mineral. It is a mixture of principally Antimony Trisulphide Sb_2S_3, and a small quantity of Antimony Trioxide Sb_2O_3.

Prepared by boiling purified antimony sulphide with solution of soda, which forms sodium antimonite, adding dilute sulphuric acid as long as a precipitate is produced, and should be kept in well-stoppered bottles, away from the light.

It is an amorphous reddish brown powder, which becomes lighter on exposure to the air. Insoluble in water or alcohol, but soluble in hydrochloric acid.

It is alterative, diaphoretic and emetic. Dose: 1 to 5 grains (0.065 to 0.325 gms.) Official in compound pills of antimony.

PILULÆ ANTIMONII COMPOSITA (Compound Pills of Antimony). Plummer's Pills, Compound Calomel Pills, Compound Pills of Subchloride of Mercury.

Contain sulphurated antimony, calomel, guaiac, and castor oil. Dose: 1 to 2 pills.

Arsenic Compounds.

ARSENIC (not arsenous acid) is a steel grey metallic element obtained by heating arsenous acid with charcoal, or on a large scale by heating arsenical pyrites in earthen tubes, when the metal sublimes and sulphide of iron is left. Arsenic forms two oxides with oxygen, namely: arsenous oxide, As_2O_3, the official arsenous acid, and arsenic oxide As_2O_5.

ARSENI IODIDUM (Arsenic Iodide) AsI_3.

Prepared by the direct combination of arsenic with iodine, or by evaporating to dryness an aqueous solution of arsenous acid and one of hydriodic acid, and should be kept in glass-stoppered bottles, in a cool place, protected from light.

It gradually loses iodine on exposure to the air. Soluble in water and alcohol. Used externally in skin diseases. Dose: $\frac{1}{8}$ gr. (0.008 gm.).

Barium Compounds.

BARIUM is the metal present in baryta (barium oxide), which is obtained from the native carbonate or sulphide. It is difficultly fusible, of a silver gray color, decomposing water readily, and is considerably heavier than sulphuric acid.

BARII DIOXIDUM (Barium Dioxide) BaO_2, Barium Peroxide.

Prepared by heating barium oxide in a current of air, and should be kept in well-closed vessels. When exposed to the air it slowly attracts moisture and carbon dioxide, and is gradually decomposed. It is used in making solution hydrogen peroxide.

AQUA HYDROGENII DIOXIDI (Solution Hydrogen Dioxide) H_2O_2. Oxygenized Water, Oxygen Hydrate, Solution Hydrogen Peroxide.

Prepared by mixing barium dioxide with water, decomposing with phosphoric acid, forming hydrogen dioxide in solution and barium phosphate precipitated. The barium phosphate is filtered out and any trace of barium salt is removed by the addition of sulphuric acid, which precipitates the insoluble barium sulphate; it is then filtered through starch to prevent the fine barium sulphate from passing through the filter, and adjusted to the strength of 3% by weight of pure dioxide, equal to ten volumes of available oxygen. It has an acid reaction, owing to a little free acid that is left in it to prevent decomposition. It deteriorates by age, exposure to heat or protracted agitation. Alkalies decompose it with the formation of oxygen gas and water. It is a powerful antiseptic.

Bismuth Compounds.

BISMUTHUM (Bismuth) Bi is a crystalline metal occurring as a sulphide, rarely as an oxide, and is found principally in Saxony. It is a brittle, pulverizable, brilliant metal of a crystalline form, having a white color and reddish tint. As it occurs in commerce it is generally contaminated with other metals, principally arsenic.

The garlicky odor often given to the breath after taking bismuth salts is said to be due to traces of tellurium, and by some is claimed to be due to arsenic.

BISMUTHI CITRAS (Bismuth Citrate) $BiC_6H_5O_7$.

Prepared by boiling bismuth subnitrate with citric acid and water, adding water, collecting the precipitate, washing and drying. Permanent in the air. Insoluble in water or alcohol, but soluble in ammonia water. Used in preparing bismuth and ammonia.

BISMUTHI ET AMMONII CITRAS (Bismuth and Ammonium Citrate).

Prepared by dissolving bismuth citrate in ammonia water, filtering, evaporating on glass plates and scaling, and should be kept in small, well-stoppered bottles, away from the light.

It becomes opaque on exposure to the air. Soluble in water, and sparingly soluble in alcohol. Dose: 1 to 3 grains (0.065 to 0.2 gms.).

BISMUTHI SUBCARBONAS (Bismuth Subcarbonate). $(BiO)_2CO_3$, H_2O.

Prepared by dissolving bismuth in nitric acid and water (a portion of which oxidizes the metal with the evolution of nitrous vapors, while another portion combines with the oxide produced to form bismuth nitrate. The arsenic is also oxidized by the nitric acid, and unites with a portion of the oxidized metal to produce bismuth arseniate.) It is then diluted with water, set aside for twenty-four hours and filtered (the bismuth subarseniate being precipitated and filtered off.) The filtrate is added to a large quantity of water, and ammonia water added with constant stirring; the bismuth subnitrate is precipitated, the ammonia aiding in its thorough precipitation by uniting with the nitric acid. The precipitated subnitrate is washed, dissolved in nitric acid and water, set aside for twenty-four hours and the solution filtered. It is next added to a solution of sodium carbonate, whereby double decomposition takes place, bismuth subcarbonate being precipitated and sodium nitrate held in solution.

Permanent in the air. Insoluble in water, but soluble in nitric and hydrochloric acids with effervescence. Its uses and doses are the same as the subnitrate, except that it is more tonic.

BISMUTHI SUBNITRAS (Bismuth Subnitrate) $BiONO_3, H_2O$.

Prepared by dissolving bismuth in nitric acid and water, setting aside for twenty-four hours and filtering, after having been diluted with water, and adding to a solution of sodium carbonate (whereby most of the arsenic is retained in solution as sodium arsenate, and the insoluble carbonate precipitated) filtering, dissolving the carbonate in nitric acid and water, adding water until a permanent turbidity is produced (thus allowing any arsenic that may be present in solution to be deposited) filtering, diluting with water, adding ammonia water, which precipitates the subnitrate.

Permanent in the air. Insoluble in water and alcohol, but soluble in nitric and hydrochloric acids. Incompatible with potassium iodide (slowly forming brick-red bismuth iodide) and with alkaline carbonates.

Used in gastritis and stomach disorders. Dose: 5 to 40 grains (0.324 to 2.6 gms.).

BROMINE COMPOUND.

BROMUM (Bromine) Br. is a liquid non-metallic element, obtained from sea water and from saline springs, of a dark red color and very volatile.

It is produced largely from the salt wells in Ohio, West Virginia and Pennsylvania, by pumping out the brine, evaporating to separate chlorides, sulphates, etc., by crystallization. The mother liquor, which consists principally of magnesium bromide, is decomposed by treating with chlorine gas produced from manganese dioxide and hydrochloric acid, with the formation of magnesium chloride and bromine.

It is sparingly soluble in water, more so in alcohol and still more so in ether, chloroform and carbon disulphide.

Its use is about the same as iodine. Its antidote is ammonia largely diluted with water. It unites with the bases to form bromide.

Calcium Compounds.

CALCIUM (Calcium) Ca. is a pale yellow metallic element, existing in nature as a carbonate in the form of chalk, limestone, marble, oyster shells, etc.

CALX (Lime) CaO. Burned Lime, Quicklime.

Prepared by burning limestone, marble, oyster shells; carbon dioxide and water being expelled, and should be kept in well closed vessels in a dry place.

On exposure to air it gradually attracts moisture and carbon dioxide and falls to a white powder. Calcium is a metallic element, while Calx is the oxide of Calcium. Externally it is escharotic.

Official in the water, syrup and potassa with lime.

LIQUOR CALCIS (Lime Water) Solution of Lime.

Prepared by saturating water with Lime, and contains 0.17%.

When exposed to the air it absorbs carbon dioxide.

It is antacid, astringent and tonic. Dose: 2 to 4 fluid ounces (60 to 120 C.c.). Official in the Liniment.

SYRUPUS CALCIS (Syrup of Lime).

Prepared by triturating sugar with lime, adding boiling water, straining, diluting with an equal volume of water, filtering, evaporating and adding water to make the required quantity. It contains 6.5% of lime, which forms a saccharate with the sugar.

Dose: 20 mimins (1.25 C.c.) equivalent to a fluid ounce of Lime Water.

LINIMENTUM CALCIS (Liniment of Lime) Carron Oil.

Prepared by mixing equal volumes of lime water and linseed oil, the oil forming a soap with the lime.

It is a useful application in burns and scalds.

CALCII BROMIDUM (Calcium Bromide) $CaBr_2$.

Prepared by acting on sulphur with bromine, forming sulphur bromine, and acting on this with milk of lime, forming calcium bromide and sulphate, which may be easily separated. Also prepared by adding precipitated calcium carbonate in excess to hydrobromic acid, filtering, evaporating to dryness and granulating, and should be kept in well stoppered bottles.

It is deliquescent on exposure to the air. Soluble in water and alcohol. Sometimes used as a substitute for potassium bromide.

Dose: ½ to 2 drachms (1.95 to 7.8 gms.).

CALCII CARBONAS PRECIPITATUS (Precipitated Calcium Carbonate) $CaCO_3$. Precipitated Chalk.

Prepared by precipitating a solution of calcium chloride with one of sodium carbonate.

Permanent in the air. Insoluble in water or alcohol. Soluble in nitric, hydrochloric, and acetic acids with effervescence. Used as an antacid in tooth powder and in making prepared chalk.

CRETA PREPARATA (Prepared Chalk).

Chalk freed from most of its impurities by elutriation, and moulded in conical masses Whiting, a cheap form of chalk, is used for polishing. Prepared Chalk is permanent in the air, and insoluble in water or alcohol. It is the only form of chalk used in medicine, being antacid. Official in mercury with chalk, compound chalk powder, and troche.

PULVIS CRETÆ COMPOSITUS (Compound Chalk Powder).

Contains prepared chalk, powdered acacia, powdered sugar. Official in chalk mixture.

MISTURA CRETÆ (Chalk Mixture).

Contains compound chalk powder, cinnamon water and water.

It should be freshly prepared when wanted for use. It is a convenient form of administering chalk in looseness of the bowels. Dose : 1 tablespoonful (15 C.c.).

TROCHISCI CRETÆ (Troches of Chalk).

Contain prepared chalk, powdered acacia, spirit of nutmeg, sugar and water.

Each troche contains 4 grains of prepared chalk. Used as a gentle astringent.

CALCII CHLORIDUM (Calcium Chloride) $CaCl_2$.

Prepared by saturating hydrochloric acid with marble or chalk, evaporating to dryness and heating to redness. It should be kept in well-stoppered bottles.

Very deliquescent. Soluble in water and alcohol.

CALCII HYPOPHOSPHIS (Calcium Hypophosphite) $Ca(PH_2O_2)_2$.

Prepared by boiling milk of lime with phosphorus until the phosphoretted hydrogen gas, which is inflammable, is driven off, filtering, concentrating and cyrstallizing.

Permanent in the air. Soluble in water and insoluble in alcohol. Incompatible with the soluble salts of mercury, copper and silver, which are reduced.

Used for the same purpose as the other hypophosphites. Dose : 10 to 30 grains (0.65 to 1.95 gm.) Official in syrup of hypophosphites.

CALCII PHOSPHAS PRECIPITATUS (Prepared Calcium Phosphate) $Ca_3(PO_4)_2$.

Prepared by dissolving bone ash (impure calcium phosphate) in hydrochloric acid and water, and precipitating the phosphate with ammonia water, chloride of ammonium remaining in solution.

Permanent in the air. Insoluble in water and alcohol. It has replaced magnesium carbonate and absorbent cotton in the preparation of the medicated waters.

Official in antimonial powder. Dose : 10 to 30 grs. (0.65 to 1.95 gm.).

LECTURE NO. 6.

SYRUPUS CALCII LACTOPHOSPHATIS (Syrup of Lactophosphate of Calcium).

Prepared by dissolving precipitated calcium carbonate in lactic acid, adding phosphoric acid, and mixing with orange flower water in which sugar has been dissolved.

Dose: 2 to 4 fluid drachms (7.5 to 15 C.c.).

CALCII SULPHAS EXSICCATUS (Dried Calcium Sulphate) $CaSO_4$. Sulphate of Lime, Gypsum, Plaster of Paris.

It exists native as gypsum, and when heated loses its water of crystallization and becomes Plaster of Paris. It contains about 95% by weight of calcium sulphate and about 5% of water. When mixed with half its weight of water it forms a smooth, cohesive paste which rapidly hardens.

CARBON COMPOUNDS.

CARBON (C.) is one of the most important elements that exist, and is found in all organic compounds, and exists in nature in the form of charcoal, diamond, coal, plumbago, etc. Combined with oxygen it forms carbon dioxide or carbonic acid gas, and carbon monoxide. In its uncrystallized state it is an insoluble, infusible solid, generally of a black color, and without taste or smell. It is unalterable and indestructible, and has great power in resisting and correcting putrefaction in other bodies. It possesses the property of absorbing the color and odorous principles of most liquids.

CARBON DIOXIDE (CO_2) is a colorless, odorless gas, heavier than ordinary air and having a slightly acid taste, obtained by acting on carbonates with acids, or by heating carbonates.

CARBONIC ACID (H_2CO_3) is prepared by bringing carbonic acid in contact with water. Soda water is a solution of carbonic acid gas in water.

CEREUM COMPOUNDS.

CERII OXALAS (Cerium Oxalate) $Ce_2(C_2O_4)_3$, 9 H_2O.

Prepared by adding a solution of ammonium oxalate to a soluble salt of cereum. It exists native as Cerite.

Permanent in the air. Insoluble in water or alcohol. Soluble in diluted sulphuric or hydrochloric acids.

Its action is similar to that of bismuth subnitrate. Dose: 1 gr. (0.065 gm.).

CHLORINE COMPOUNDS.

CHLORINE (Cl.) is a gaseous element prepared by the action of hydrochloric acid upon manganese dioxide, with the aid of heat, and exists in nature in the form of sodium chloride, called rock salt or sea salt.

It is a greenish yellow gas, with a suffocating odor. It is one of the best disinfectants and is used considerably as a bleach for organic coloring principles in the presence of water.

AQUA CHLORI (Chlorine Water).

Prepared by saturating water with chlorine gas, and contains 0.4%.

Incompatible with alkalies, silver and lead salts, infusions, tannin, tinctures, emulsions, and liberates bromine from any of its salts. When prescribed in mixtures it should be added last. It should be prepared fresh when wanted for use. It is stimulant and antiseptic. Dose : 1 to 4 fluid drachms (3.75 to 15 C.c.).

CALX CHLORATA (Chlorinated Lime) $CaOCl_2$. Hypochlorite of Lime, Oxymuriate of Lime, Bleaching Powder. Often improperly called Chloride of Lime.

Prepared by passing chlorine gas into slaked lime as long as it is absorbed, and should contain not less than 35%.

It is a powerful oxidizing agent, converting sugar, starch, cotton, linen and similar substances into formic acid which unites with the lime.

Incompatible with the mineral acids, carbonic acid, and the alkaline carbonates. Used in making solution of chlorinated soda, and also as a dessicant.

LIQUOR SODÆ CHLORATÆ (Solution of Chlorinated Soda). Labarraque's Solution.

Prepared by mixing together a hot solution of sodium carbonate and a cold solution of chlorinated lime. The reason for using a hot solution of sodium carbonate is to make the precipitated calcium carbonate as dense as possible so that it may rapidly settle. It should contain at least 2.6% of available chlorine.

Incompatible with many metallic salts, and with the iodides and bromides, causing the liberation of free iodine and bromine.

Used as a stimulant, antiseptic and disinfectant. Dose : 30 to 60 drops (1.9 to 3.9 C.c.).

Copper Compound.

CUPRUM (Copper) Cu. is a brilliant, reddish colored element existing native as an oxide, as a sulphide, or as a salt (principally sulphate, carbonate, arsenate and phosphate). It is obtained in the United States, Spain, Portugal and Germany.

CUPRI SULPHAS (Copper Sulphate) $CuSO_4$. Blue Vitriol, Blue Stone.

Prepared by the action of dilute sulphuric acid on waste copper and crystallizing.

Slowly efflorescent in dry air. Soluble in water, but almost insoluble in alcohol. It is stimulant, astringent and escharotic. Dose : $\frac{1}{4}$ grain (0.016 gm.). Antidote : Potassium ferrocyanide, which forms the insoluble copper ferrocyanide.

Hydrogen Compounds.

HYDROGEN (H.) is prepared by the action of sulphuric acid on zinc in the presence of water.

It is a colorless, inodorous, tasteless and combustible gas. It is one of the most important elements, in that their atomic weights are all referred to it as unity. It is the base of the acids.

WATER (H_2O) is the most important compound in existence. In its pure state it is a transparent liquid, without color, taste or odor. Its specific gravity is assumed to be unity (1). A cubic inch of it, at the temperature of 15.5"C (60°F) weighs very nearly 252.5 grains. In the metric system the weight of 1 C. c. of distilled water taken at 4°C (39.2°F) is made equal to 1 gram, the unit of weight in this system. On account of its extensive solvent powers it is more or less contaminated with foreign matters. When the foreign substances are in so small an amount as not to materially alter its taste and other sensible qualities it is called COMMON WATER. When it contains inconsiderable quantities of impurities, and when used in washing, forms a lather with soap, it is called SOFT WATER. When it contains calcareous or magnesia salts, or other impurities, through which it curdles soap and is unfit for domestic purposes, it is called HARD WATER. Common water is divided into varieties according to its source, namely: Rain or Snow water, Spring water, River water, Well Water, Lake water and Marsh water.

AQUA DISTILLATA (Distilled Water). Prepared by distilling water and throwing away the first ten and the last ten per cent., and saving the intermediate eighty per cent. as finished product.

The first part and the last part may contain carbonic acid and other volatile impurities.

Aqua Hydrogenii Dioxodi: See Barium Compounds.

Gold Compounds.

AURI ET SODII CHLORIDUM (Gold and Sodium Chloride).

Prepared by mixing together equal parts of sodium chloride and gold chloride (which is prepared by dissolving gold in nitrohydrochloric acid and evaporating to dryness).

Slightly deliquescent in damp air. Soluble in water.

It is alterative. Dose $\frac{1}{4}$ to $\frac{1}{12}$ of a grain (0.015 to 0.005 gm.).

Iodine Compounds.

IODUM (Iodine) I.

A non-metallic element obtained from the ashes of seaweed and from mineral iodides and iodates. Also prepared by decomposing crude sodium iodide with manganese dioxide and sulphuric acid and subliming. It should be kept in glass-stoppered bottles, in a cool place.

Soluble in alcohol and very little soluble in water.

Its solubility is increased in water by the addition of sodium chloride, ammonium nitrate, potassium iodide. It volatilizes slowly at ordinary temperature. It may be detected by the addition of starch test solution, producing a blue color. Sometimes adulterated with charcoal, mineral coal, plumbago and manganese dioxide. Incompatible with alkaloids.

Used principally as an irritant and alterative. Dose: $\frac{1}{4}$ to 1 grain (0.015 to 0.065 gm.).

Official in the tincture, ointment and compound solution.

LIQUOR IODI COMPOSITUS (Compound Solution of Iodine) Lugoll's Solution, Solution of Iodine.

Prepared by dissolving iodine with the aid of potassium iodide in water, and contains 5%. It should be kept in glass-stoppered bottles.

It should never be compounded in any prescription with an alkaloid, as iodine unites with the alkaloids, forming double salts. It is a convenient form for the administration of Iodine. Dose: 5 minims (0.3 C.c.) containing about $\frac{1}{4}$ of a grain.

TINCTURA IODI (Tincture of Iodine).

An alcoholic solution of iodine containing 7%. It is well to prepare the tincture in small quantities, because the iodine reacts with the alcohol in sunlight, causing chemical change. The iodine should be thoroughly dried before using. Dose: 5 to 15 drops (0.31 to 0.93 C.c.).

UNGUENTUM IODI (Ointment of Iodine).

Prepared by rubbing iodine and potassium iodide with water and then with benzoinated lard. The object of the potassium iodide and water is to bring the iodine into a state in which it can be incorporated with the lard. It is useful as an external application in swellings, etc.

SYRUPUS ACIDI HYDRIODICI (Syrup of Hydriodic Acid).

Contains about 1% by weight of absolute acid (HI). Prepared by dissolving potassium iodide and potassium hypophosphite in water, and tartaric acid in diluted alcohol, mixing the solutions, and after standing for a short time filtering into syrup. The object of the dilute alcohol is to aid in the precipitation of the potassium bitartrate, and the potassium hypophosphite is used to prevent the formation of free iodine.

Incompatible with potassium chlorate, mineral acids and the salts of the metals on account of liberating free iodine.

It has the same therapeutical properties as iodine. Dose: 20 to 40 minims (1.25 to 2.5 C.c.).

IRON COMPOUNDS.

FERRUM (Iron) Fe. is in the form of fine, bright and non-elastic wire. It exists in nature in many combinations, principally in the form of sulphide (iron pyrites). It is a hard, malleable, ductile and tenacious metal of a grayish-white color and fibrous texture.

At a red heat its surface is converted into black oxide and at common temperatures by the action of air and moisture it is converted into a yellowish brown hydrated sesquioxide (iron rust). It is detected even in small quantities, by oxidizing with nitric acid, and adding potassium ferrocyanide or tincture of galls, the former producing a deep blue color, and the latter a black. Iron filings should not be used in pharmacy on account of the dirt and grease that they usually contain.

FERRUM REDUCTUM (Reduced Iron), Powder of Iron, Iron reduced by Hydrogen, Iron by Hydrogen.

Prepared by passing hydrogen gas through ferric oxide heated to redness in an iron tube. The hydrogen gas unites with the oxygen of the ferric oxide to form water and the metallic iron in powder is left.

On exposure to the air it gradually oxidizes, and on this account should be kept in well-stoppered bottles in a cool place. It is one of the best chalybeate tonics. Dose: 3 to 6 grs. (0.20 to 0.40 gms.).

LIQUOR FERRI ACETATIS (Solution of Ferric Acetate), Solution of Peracetate of Iron. Contains about 31% of the anhydrous salt corresponding to about 7.5% of metallic iron.

Prepared by dissolving ferric hydrate (prepared by precipitating a solution of iron tersulphate with ammonia water) in glaciale acetic acid.

It is chalybeate. Dose: 2 to 10 minims (0.12 to 0.60 C.c.).

FERRI CARBONAS SACCHARATUS (Saccharated Ferrous Carbonate) $FeCO_3$.

Prepared by precipitating a solution of ferrous sulphate with one of sodium bicarbonate, collecting the precipitate, washing and mixing with sugar and drying. The object of the sugar is to prevent the oxidation of the iron to the ferric state. It must be kept away from the air to prevent gradual oxidation. Partially soluble in water. Soluble in hydrochloric acid. It is an excellent chalybeate. Dose: 5 to 30 grains (0.33 to 1.95 gm.).

FERRI CHLORIDUM (Ferric Chloride) $Fe_2Cl_6, 12H_2O$.

Prepared by dissolving iron wire in hydrochloric acid and water, oxidizing the ferrous chloride with nitric acid and the aid of heat, and crystallizing.

Very deliquescent in moist air. Soluble in water and alcohol. Used in the form of solution or tincture.

MASSA FERRI CARBONATIS (Mass of Ferrous Carbonate), Vallet's Mass, Pill Carbonate of Iron, Ferruginous Mass.

Prepared by precipitating a solution of ferrous sulphate with one of sodium carbonate, washing the precipitate with a mixture of syrup and water (to prevent the absorption of oxygen and the loss of carbonic acid gas), draining, mixing with honey and sugar and evaporating to a pilular consistence.

Contains about 35% of ferrous carbonate. It becomes black on exposure. It is an excellent chalybeate. Dose: 3 to 5 grains (0.20 to 0.33 gm.).

MISTURA FERRI COMPOSITA (Compound Iron Mixture), Griffeth's Mixture.

Prepared by double decomposition between ferrous sulphate and potassium carbonate, myrrh being used as a protecting agent, and flavored with spirit of lavender and rose water. Contains about 2% of ferrous carbonate. It is a good tonic in debility of the digestive organs. Dose: 1 to 2 fluid ounces (30 to 60 C.c.).

PILULÆ FERRI CARBONATIS (Pills of Ferrous Carbonate). Blaud's Pills, Ferruginous Pills, Chalybeate Pills, Iron Pills.

Contains ferrous sulphate mixed with sugar, then mixed with potassium carbonate, to which a little glycerin and water have been added: triturating and beating until the mass assumes a greenish color, and when the reaction appears to have ceased, adding tragacanth, marshmallow, and a little water, if necessary to make the mass.

They should be freshly prepared when wanted for use. Dose: 2 to 6 pills.

LIQUOR FERRI CHLORIDI (Solution of Ferric Chloride), Solution of Chloride of Iron, Strong Solution of Perchloride of Iron.

Contains 37.8% of the anhydrous salt, corresponding to 62.9% of the crystallized salt, or to about 13% of metallic iron.

Prepared by dissolving iron wire in hydrochloric acid and water, and oxidizing to the ferric state with nitric acid and heat. If the solution should have a blackish color, instead of a clear ruby red, it is due to the presence of a nitro compound composed of a portion of ferrous chloride and nitrogen dioxide. As this is easily decomposed it is only necessary to heat the liquid and add a few drops of nitric acid, when the blackish color will soon disappear, nitrogen dioxide being liberated and a clear ruby red solution remaining.

It is a powerful styptic. Dose: 2 to 10 minims (0.12 to 0.6 C. c.) Official in the tincture.

TINCTURA FERRI CHLORIDI (Tincture of Ferric Chloride), Tincture Muriate of Iron, Tincture of Perchloride of Iron, Tincture of Sesquichloride of Iron.

A hydro-alcoholic solution of ferric-chloride containing 13.6% of the anhydrous salt, corresponding to about 4.7% of metallic iron.

Prepared by mixing together 25 C.c. of the solution and 75 C.c. of alcohol.

The U. S. P. directs that it be kept at least three months before using, to allow the development of ethyl chloride and similar ethereal compounds, which impart an agreeable odor to the preparation.

Incompatible with alkalies (alkali earths or carbonates, with production of the hydrate or oxide) likewise with preparations containing tannic acid (inky mixtures), also with mercurous salts (forming mercuric compounds) and with mucilage of acacia it forms a jelly.

It is chalybeate. Dose: 10 to 30 minims (0.6 to 1.9 C.c.).

FERRI CITRAS (Ferric Citrate), $Fe_2(C_6H_5O_7)_2, 6H_2O$.

Prepared by the evaporation of solution of ferric citrate at a temperature not exceeding 60°C, till the consistency of syrup, and spreading on glass plates and drying, and should be kept in well-stoppered bottles, away from the light.

Soluble in water, more so in hot water, but insoluble in alcohol.

It is chalybeate. Dose: 3 to 5 grains (0.2 to 0.33 gm.).

LIQUOR FERRI CITRATIS (Solution of Ferric Citrate). Contains 7.5% of metallic Iron.

Prepared by dissolving ferric hydrate in a solution of citric acid in water. It is an excellent chalybeate. Dose: 10 minims (0.6 C.c.).

LIQUOR FERRI DIALYSATUS (Solution of Dialysed Iron.) It is not official, but is used to a large extent.

Prepared by treating a solution of ferric chloride and water with ammonia, whereby ferric hydrate is precipitated, which is dissolved by agitation in a small quantity of solution of ferric chloride. The thick liquid is then placed in a dialyzer which is suspended in water and the water renewed until the diffusate gives no trace of hydrochloric acid.

It is a solution of ferric oxychloride in water.

It is a ferruginous tonic, free from astringency. Dose: 20 to 60 minims (1.3 to 3.7 C.c.).

FERRI ET AMMONII CITRAS (Iron and Ammonium Citrate).

Prepared by mixing a solution of ferric citrate with ammonia water, evaporating at a temperature not exceeding 60°C (140°F) and scaling. It is more soluble than ferric citrate.

Deliquescent in moist air, and should be kept in well-stoppered bottles, away from the light. Soluble in water and insoluble in alcohol. Used in preparing iron and strychnine citrate, and is official in wine of citrate of iron.

Dose: 5 grains (0.33 gm.).

FERRI ET STRYCHNINÆ CITRAS (Iron and Strychnine Citrate).

Prepared by dissolving iron and ammonium citrate in water, adding citric acid and strychnine, evaporating and scaling.

Contains 1% of strychnine. Deliquescent in damp air, and should be kept in well-stoppered bottles, away from the light. Soluble in water and partially so in alcohol.

It is tonic. Dose: 3 to 5 grains (0.20 to 0.33 gm.).

VINUM FERRI CITRATIS (Wine of Citrate of Iron).

Contains citrate of iron and ammonium, tincture of sweet orange peel, syrup and white wine. It is chalybeate. Dose: 1 fluid drachm (3.75 C.c.), which contains about $2\frac{1}{4}$ grains of the iron salt.

FERRI ET AMMONII SULPHAS (Iron and Ammonium Sulphate). Ammonio-ferric Alum, Ammonio-ferric Sulphate.

Prepared by dissolving ammonium sulphate in a hot solution of iron tersulphate and crystallizing.

Efflorescent on exposure to the air. Soluble in water and insoluble in alcohol. It is styptic. Dose: 5 to 10 grains (0.33 to 0.65 gm.)

FERRI ET AMMONII TARTRAS (Iron and Ammonium Tartrate), Ammonio-ferric Tartrate, Ammonio-tartrate of Iron.

Prepared by dissolving ferric hydrate in a solution of acid ammonium tartrate, evaporating and scaling. Slightly deliquescent in the air, and should be kept in well-stoppered bottles, away from the light. Soluble in water and insoluble in alcohol. It is a mild chalybeate. Dose: 10 to 30 grains (0.65 to 1.95 gm.).

FERRI ET POTASSII TARTRAS (Iron and Potassium Tartrate), Potassio-ferric Tartrate, Tartrated Iron.

Prepared by dissolving ferric hydrate in solution potassium bi tartrate, evaporating and scaling.

Slightly deliquescent in the air, and should be kept in well-stoppered bottles, away from the light. Soluble in water and insoluble in alcohol. It is a mild, agreeable chalybeate. Dose: 5 to 10 grains (0 33 to 0.65 gm.).

FERRI ET QUININÆ CITRAS (Iron and Quinine Citrate). Contains about 11.5% of quinine and 14.5% of metallic iron.

Prepared by dissolving quinine, citric acid and ferric citrate in water, evaporating and scaling.

Slowly deliquescent in damp air, and should-be kept in well-stoppered bottles, away from the light. Soluble in water, and partially soluble in alcohol.

It is a tonic. Dose: 5 or 6 grains (0.33 or 0.4 gm.).

FERRI ET QUININÆ CITRAS SOLUBILIS (Soluble Iron and Quinine Citrate).

Prepared by dissolving iron citrate in distilled water, heating, adding quinine and citric acid, previously triturated with water, adding ammonia water, evaporating and scaling.

Deliquescent in damp air and should be kept in well-stoppered bottles, away from the light. Soluble in water and partially soluble in alcohol.

Official in bitter wine of iron.

VINUM FERRI AMARUM (Bitter Wine of Iron). Contains soluble citrate of iron and quinine, tincture of sweet orange peel, syrup and white wine. It is a mild, ferruginous tonic. Dose: 2 to 4 fluid drachms (7.5 to 15 C.c.).

LECTURE NO. 7.

FERRI HYPOPHOSPHIS (Ferric Hypophosphite) $Fe_2(PH_2O_2)_6$.
Prepared by acting on a solution of ferrous sulphate with one of calcium hypophosphite, whereby calcium sulphate is precipitated and ferrous hypophosphite is converted into the ferric state by evaporation. Permanent in the air. Slightly soluble in water, but more so in the presence of hypophosphorus acid. Dose: 5 to 10 grs. (0.32 to 0.64 gm.).

FERRI IODIDUM SACCHARATUM (Saccharated Ferrous Iodide) FeI_2.
Prepared by the action of iodine on iron in the presence of water, allowing the mixture to stand until it has obtained a green color, mixing with sugar of milk, drying and mixing with more sugar of milk and some reduced iron.

On standing it is liable to oxidize with the liberation of free iodine, which unites with the reduced iron.

Soluble in water and insoluble in alcohol. It is chalybeate. Dose: 2 to 5 grains (0.13 to 0.33 gm.).

PILULÆ FERRI IODIDI (Pills of Ferrous Iodide), Blanchard's Pills.

Contain reduced iron, iodine, licorice, extract of licorice, sugar, acacia, and are coated with balsam of tolu, dissolved in ether.

Each pill contains about 1 grain of ferrous iodine, and about one-fifth of a grain of reduced iron.

FERRI LACTAS (Ferrous Lactate) $Fe(C_3H_5O_3)_2, 3H_2O$.
Prepared by the action of lactic acid on iron wire, and should be kept in well-stoppered bottles.

Soluble in water and insoluble in alcohol.

It has the same properties as the ferruginous preparations. Dose: 1 to 5 grains (0.065 to 0.325 gm.). Official in syrup hypophosphites with iron.

FERRI OXIDUM HYDRATUM (Hydrated Oxide of Iron), $Fe_2(OH)_6$. Ferric Hydrate, Hydrous Peroxide of Iron, Hydrated Sesquioxide of Iron, Moist Peroxide of Iron.

Prepared by precipitating a solution of tersulphate of iron with ammonia water.

It is a brownish red magna, wholly soluble in hydrochloric acid, without effervescence.

Official in the troche and plaster.

EMPLASTRUM FERRI (Iron Plaster), Strengthening Plaster, Chalybeate Plaster.

Prepared by adding ferric hydrate to a mixture of olive oil, burgundy pitch and lead plaster.

TROCHISCI FERRI (Troches of Iron). Contain ferric hydrate, vanilla, sugar, and mucilage of tragacanth. Each lozenge contains about 5 grains ferric hydrate.

LIQUOR FERRI NITRATIS (Solution Ferric Nitrate), Solution of Pernitrate of Iron. Contains about 6.2% of the anhydrous salt, corresponding to 1.4% of metallic iron.

Prepared by dissolving ferric hydrate in a mixture of nitric acid and water.

It is tonic and astringent.

Dose: 7 to 8 drops (0.36 to 0.42 C.c.).

FERRI OXIDUM HYDRATUM CUM MAGNESIA (Ferric Hydrate with Magnesia), Arsenic Antidote, Antidote to Arsenous Acid.

Prepared by mixing a diluted solution of tersulphate of iron with magnesia and water. These two solutions should be kept all ready for use. Ferric hydrate is produced when the magnesia mixture is added to the diluted solution of iron tersulphate, the acidity being prevented by using an excess of magnesia. It acts as an antidote by uniting with the arsenous acid to form arsenite of iron, which is insoluble.

FERRI PHOSPHAS SOLUBILIS (Soluble Ferric Phosphate) $Fe_2(PO_4)_2$.

Prepared by dissolving sodium phosphate in a solution of ferric citrate, evaporating and scaling. Should be kept in dark amber, well-stoppered bottles.

Permanent in the air, when not exposed to the light, but on exposure it becomes dark and discolored.

Official in syrup of phosphate of iron, quinine and strychnine.

SYRUPUS FERRI, QUININÆ ET STRYCHNINÆ PHOSPHATUM (Syrup of Phosphates of Iron, Quinine and Strychnine), Easton's Syrup, Aitkin's Syrup, Syrup of Three Phosphates, Syrup of Triple Phosphates.

Contains soluble ferric phosphate, quinine sulphate, strychnine, phosphoric acid, glycerin, syrup and water. Dose: 1 teaspoonful (3.7 C.c.).

FERRI PYROPHOSPHAS SOLUBILIS (Soluble Ferric Pyrophosphate) $Fe_4(P_2O_7)_3$.

Prepared the same as the soluble phosphate, except using pyrophosphate of sodium in the place of phosphate, and should be kept in amber-colored, well-stoppered bottles. Soluble in water, but insoluble in alcohol.

• It is a good chalybeate. Dose: 2 to 5 grains (0.13 to 0.33 gm.).

FERRI SULPHAS (Ferrous Sulphate) $FeSO_4, 7H_2O$, Sulphate of Iron, Green Vitriol.

Prepared by the action of sulphuric acid on iron wire in the presence of water, evaporating and crystallizing. Should be kept in well-stoppered bottles.

Efflorescent in dry air, but on exposure to moist air it rapidly absorbs oxygen and becomes coated with brownish, yellowish basic ferric sulphate. Soluble in water and insoluble in alcohol. Incompatible with lime water, calcium and barium chlorides, alkalies and their carbonates, soap, silver nitrate, sodium borate and phosphate, lead acetate and subacetate. It is used in making the dried and granulated sulphate of iron. It is an astringent chalybeate. Dose: 1 to 2 grains (0.065 to 0.130 gm.).

FERRI SULPHAS EXSICCATUS (Dried Ferrous Sulphate).

Prepared by allowing ferrous sulphate to effloresce in the air, heating and powdering. Should be kept in dry, well stoppered bottles. Three grains of it are equal to about five grains of the sulphate. It is used principally in making pills on account of its reduced bulk and loss of water.

Official in pills of aloes and iron.

PILULÆ ALOES ET FERRI (Pills of Aloes and Iron).

Contain purfied aloes, dried sulphate of iron, aromatic powder and confection of rose.

They are laxative. Dose: 1 to 3 Pills.

FERRI SULPHAS GRANULATUS (Granulated Ferrous Sulphate).

Prepared by dissolving ferrous sulphate in water heated to boiling, adding dilute sulphuric acid, filtering while hot, evaporating, cooling quickly with constant stirring, pouring alcohol upon it in a funnel and quickly drying. Keep in dry, well-stoppered bottles.

The object of heat with constant stirring is to granulate the salt, and the alcohol washes out impurities and facilitates rapid drying of the granular powder.

It is less liable to oxidation than the sulphate.

FERRI VALERIANAS (Ferric Valerianate) $Fe_2(C_5H_9O_2)_6$.

Prepared by precipitating a diluted solution of ferric sulphate with one of sodium valerianate, collecting, washing and drying, and should be kept in small, well-stoppered bottles, in a cool and dark place.

Permanent in dry air. Soluble in alcohol but insoluble in water. It is a chalybeate tonic. Dose: 2 to 5 grains (0.13 to 0.33 gm.).

LIQUOR FERRI SUBSULPHATIS (Solution of Ferric Subsulphate) $Fe_4O(SO_4)_5$, Monsell's Solution, Solution of Basic Ferric Sulphate, Solution of Subsulphate of Iron, Solution of Persulphate of Iron. It contains about 13.6% of metallic iron.

Prepared by mixing sulphuric acid and nitric acid with water, gradually adding sulphate of iron, boiling, adding more nitric acid if necessary to convert the iron into the ferric state, and boiling until it assumes a ruby red color.

It mixes with water and alcohol in all proportions.

It is a styptic. MONSELL'S SALT is obtained by heating the solution to dryness.

LIQUOR FERRI TERSULPHATIS (Solution of Ferric Sulphate) $Fe_2(SO_4)_3$, Solution of Tersulphate of Iron, Solution of Persulphate of Iron.

Prepared the same as solution of subsulphate, only less sulphate of iron and more sulphuric acid is used. It contains about 28.7 % of normal ferric sulphate, corresponding to 8% of metallic iron It is miscible with water and alcohol in all proportions. It is a powerful astringent.

SYRUPUS FERRI IODIDI (Syrup of Ferrous Iodide.) Contains about 10% by weight of ferrous iodide.

Prepared by acting on iron with iodine in the presence of water, and filtering into syrup.

It should be exposed to sunlight, if possible, to prevent the decomposition of the syrup and the liberation of free iodine.

Dose: 15 to 30 minims (0.9 to 1.9 gm.).

Lithium Compounds.

LITHII BENZOAS (Lithium Benzoate) Li $C_7H_5O_2$.

Prepared by decomposing lithium carbonate with benzoic acid. Permanent in the air. Soluble in water and alcohol.

Used as a remedy against gout. Dose: 15 to 30 grains (1 to 1.95 gm.). The presence of sodium benzoate increases its solubility in water and lessens it in alcohol.

LITHII BROMIDUM (Lithium Bromide) LiBr.

Prepared by forming lithium sulphate with lithium carbonate and sulphuric acid, and acting on the sulphate with potassium bromide. Also obtained by dissolving lithium carbonate in hydrobromic acid.

It is very deliquescent and should be kept in well-stoppered bottles. Soluble in water and alcohol. It has the same action as the other bromides. Dose: 15 to 30 grs. (1 to 1.95 gm.).

LITHII CARBONAS (Lithium Carbonate) Li_2CO_3.

It is found in certain minerals in iron mines and in numerous mineral springs.

Permanent in the air. Soluble in water and insoluble in alcohol. It has the same action as the alkaline carbonates. Dose: 5 to 15 grains (0.32 to 1 gm.).

LITHII CITRAS (Lithium Citrate) $Li_3C_6H_5O_7$.

Prepared by neutralizing lithium carbonate with citric acid, evaporating and crystallizing. Deliquescent on exposure to the air. Soluble in water and almost insoluble in alcohol.

Used for the same purpose as the carbonate. Dose: 10 to 30 grains (0.65 to 1.95 gm.).

LITHII CITRAS EFFERVESCENS (Effervescent Lithium Citrate).

It is an effervescent mixture of lithium carbonate, citric acid, sodium bicarbonate and sugar, and should be kept in well-stoppered bottles.

Dose : 1 teaspoonful.

LITHII SALICYLAS (Lithium Salicylate) Li $C_7H_5O_3$.

Prepared by neutralizing lithium carbonate with salicylic acid. Deliquescent on exposure to the air. Soluble in water and alcohol.

Used in cases of gout and rheumatism. Dose: 20 to 40 grains (1.3 to 2.6 gm.).

Lead Compounds.

PLUMBUM (Lead) Pb, is a soft bluish gray, malleable metallic element existing native principally as an oxide or as a sulphide (galena).

PLUMBI ACETAS (Lead Acetate) $Pb(C_2H_3O_2)_2, 3H_2O$. Sugar of Lead.

Prepared by digesting lead oxide in acetic acid, evaporating and crystallizing. It should be kept in well-stoppered bottles.

Efflorescent on exposure to air, and absorbs carbon dioxide forming lead carbonate. Soluble in water and alcohol. It is decomposed by all acids, and by those soluble salts, the acids of which produce with lead protoxide, insoluble or sparingly soluble compounds. Acids of this character are citric, sulphuric, hydrochloric and tartaric. It is decomposed by lime water, ammonia, potassa and soda. The dispenser should select the large crystalline masses. The commercial acetate sometimes contains lead sulphate and carbonate. It is decomposed by hard water on account of the calcium sulphate and common salt, which such water usually contains.

Antidotes : Soluble sulphates, salt, soap or an alkali.

Used in preparing solution subacetate of lead. It is astringent and sedative. Dose : 1 to 3 grains (0.065 to 2.20 gm.).

LIQUOR PLUMBI SUBACETATIS (Solution of Lead Subacetate).

Prepared by boiling lead acetate with lead oxide and water and contains 25% of lead subacetate.

On exposure to the air it absorbs carbon dioxide, which causes the formation of a white precipitate of lead carbonate. Bottles in which the solution is kept and are coated with lead carbonate, can be easily cleaned by the addition of a little sulphuric acid. Solutions of gum, tannin, most vegetable coloring matters and many animal substances, particularly albumen, produce with it precipitates consisting of the substance added and lead oxide. It is used in preparing diluted solution subacetate of lead, and the cerate. It is astringent and sedative and is used externally.

CERATUM PLUMBI SUBACETATIS (Cerate of Lead Subacetate) Goulard's Cerate. Contains solution of subacetate of lead and camphor cerate. · Used principally in cases of burns, scalds, etc.

LIQUOR PLUMBI SUBACETATIS DILUTUS (Diluted Solution of Lead Subacetate). Lead water.

Prepared by mixing 30 C.c. of solution of lead subacetate with 970 C.c. of water. Keep in well-stoppered bottles.

It is a cloudy solution owing to the formation of the carbonate from the carbon dioxide in the water.

PLUMBI CARBONAS (Lead Carbonate) $(PbCO_3)_2 Pb(OH)_2$, White Lead, Flake White.

Prepared by passing a stream of carbonic acid gas through a solution of lead subacetate.

Permanent in the air. Insoluble in water and alcohol. It is astringent and sedative. Official in the ointment.

UNGUENTUM PLUMBI CARBONATIS (Ointment of Lead Carbonate). Contains 10% of lead carbonate mixed with benzoinated lard.

Used for blistered and excoriated surfaces.

PLUMBI IODIDUM (Lead Iodide) PbI_2.

Prepared by precipitating a solution of lead nitrate with one of potassium iodide, washing and drying.

Permanent in the air. Slightly soluble in water and almost insoluble in alcohol. Soluble in solutions of fixed alkalies, concentrated solutions of acetates of the alkalies, of potassium iodide, of sodium hyposulphite, and in a hot solution of ammonium chloride.

It should be kept away from the light.

UNGUENTUM PLUMBI IODIDI (Ointment of lead Iodide).

Contains 10% of lead iodide mixed with benzoinated lard.

PLUMBI NITRAS (Lead Nitrate) $Pb(NO_3)_2$.

Prepared by acting on lead oxide with nitric acid.

Permanent in the air. Soluble in the water and almost insoluble in alcohol.

Its uses are the same as the other lead salts.

PLUMBI OXIDUM (Lead Oxide) PbO., Litharge, Protoxide of Lead.

Obtained as a by-product in the extraction of silver from some of its ores, principally argentiferous galena.

On exposure to the air it slowly absorbs moisture and carbon dioxide. When heated in contact with charcoal it is reduced to metallic lead. Almost insoluble in water and insoluble in alcohol. Soluble in acetate acid, or dilute nitric acid, and in warm solutions of fixed alkalies.

It should not be confounded with red lead which is a higher oxide Pb_3O_4. Red Lead may be obtained by heating litharge and for this reason, in preparing litharge, care must be used in heating or the litharge may contain red lead, which would render it useless in the preparation of lead plaster.

Official in lead plaster and is used in making solution of subacetate of lead.

EMPLASTRUM PLUMBI (Lead Plaster) Diachylon Plaster.

Prepared by boiling together, litharge, olive oil and water.

The oil consists of oleate of glyceryl with a little palmitate. These are decomposed in contact with water, and unite with the litharge to form the plaster, which is chemically oleo-palmitate of lead and glycerin is formed as a by-product.

Official in diachylon ointment.

UNGUENTUM DIACHYLON (Diachylon Ointment.)

Contains lead plaster, olive oil, and oil of lavender flowers. Used in skin diseases.

Magnesium Compounds.

MAGNESIUM (Magnesium) Mg, is a white metallic element, and may be obtained from the chloride by the action of sodium. It burns readily in the air, and the white heat to which the particles of Magnesia (MgO) produced are exposed, emits a dazzling light. It exists native as a double carbonate of magnesia and calcium (Dolomite), and in a fairly pure state as Magnesite.

MAGNESIA (Magnesia) MgO. Light Magnesia, Calcined Magnesia, Magnesia Usta.

Prepared by exposing light magnesium carbonate to a red heat. Permanent in the air. Insoluble in water or alcohol.

It is liable to contain as impurities, magnesium carbonate, silica, alumina, etc. Should be kept in well-closed vessels.

Almost insoluble in water, and insoluble in alcohol, soluble in dilute acids.

If Magnesia contains a soluble sulphate or carbonate, barium chloride will show it by precipitating with water digested on the magnesia.

It enters into hydrated oxide of iron with magnesia and compound rhubarb powder. It is antacid and laxative. Dose: 10 to 30 grains (0.65 to 1.95 gm.).

MAGNESIA PONDEROSA (Heavy Magnesia) MgO.

Prepared the same as the light, only using the heavy carbonate. It possesses only one-fourth the bulk of the light, and is therefore much more convenient for internal use.

It differs from magnesia by not readily uniting with water to form a gelatinous hydrate.

MAGNESII CARBONAS (Magnesium Carbonate) $(MgCO_3)_4$ $Mg(OH)_2$, $5H_2O$.

It is found native as a mineral called magnesite. The official carbonate is a mixture of carbonate and hydrate.

The heavy carbonate is prepared by dissolving sodium carbonate and magnesium sulphate separately in boiling water and mixing, the heavy carbonate precipitating. The light carbonate is prepared the same, only using cold water.

Permanent in the air. Soluble in dilute acid, but insoluble in water or alcohol. It is decomposed by strong heat, by all acids, potassa, soda, lime, barium and strontium oxides, and by acidulous and metallic salts.

Used in making solution citrate of magnesium, and effervescent citrate of magnesium.

It is antacid. Dose: ½ to 2 drachms (1.95 to 7.8 gm.).

LIQUOR MAGNESII CITRATIS (Solution of Magnesium Citrate).

Prepared by dissolving magnesium carbonate in citric acid and water, filtering, adding syrup of citric acid and water, placing in twelve ounce bottles, adding potassium bicarbonate (39 grains to each bottle) and tightly corking.

It is a cooling purgative. Dose, as a purgative: 12 ounces (360 C.c). As a laxative: 6 ounces (180 C.c.).

MAGNESII CITRAS EFFERVESCENS (Effervescent Magnesium Citrate.).

Prepared by mixing magnesium carbonate with citric acid and water, so as to form a mass or thick paste, drying at a temperature not exceeding 30°C (86°F) and reducing to a fine powder, mixing with sugar, sodium bicarbonate and more citric acid, dampening the mixture with alcohol to form a mass, and rubbing through a No. 6 tinned iron sieve, drying and reducing to a coarse powder.

Deliquescent on exposure to air. Soluble in water and insoluble in alcohol.

Dose: 1 to 3 teaspoonfuls.

MAGNESII SULPHAS (Magnesium Sulphate) $MgSO_4$, $7H_2O$. Epsom Salt.

It is called Epsom Salt because it originally was obtained by evaporating the waters of certain springs in Epsom, England. Also, obtained from Kieserite, an impure sulphate, and from Dolomite, which is a double carbonate of magnesium and calcium.

Efflorescent in the air. Soluble in water and insoluble in alcohol. It is decomposed by potassa, soda and their carbonates, by lime, barium and strontium oxides and their soluble salts. Official in compound infusion of senna.

It is a cathartic. Dose: 1 ounce (31.1 gm.).

Manganese Compounds.

MANGANI DIOXIDUM (Manganese Dioxide) MnO_2, Black Oxide of Manganese, Peroxide of Manganese.

It exists native and should contain at least 66% of pure dioxide. Distinguished from antimony sulphide by its infusibility, and causing the evolution of chlorine on being heated with hydrochloric acid.

Permanent in the air. Insoluble in water or alcohol.

It is tonic and alterative. Dose: 3 to 20 grains (0.2 to 1.3 gm.).

LECTURE NO. 8.

MANGANI SULPHAS (Manganese Sulphate) $MnSO_4, 4H_2O$.

Prepared by heating the native black oxide with sulphuric acid, and should be kept in well-stoppered bottles. Slightly efflorescent in the air. Soluble in water, but insoluble in alcohol.

It is purgative and tonic. Dose, as a purgative: 1 to 2 drachms (3.9 to 7.5 gm.) As a tonic: 5 to 20 grains (0.33 to 1.3 gm.).

MERCURY COMPOUNDS.

HYDRARGYRUM (Mercury) Hg. Quicksilver.

Obtained from the native sulphide called cinnabar by subliming with lime, forming sulphide and sulphate of calcium and mercury sublimes over.

It is a liquid at ordinary temperatures. Its specific gravity is 13.5584. Insoluble in water or alcohol. Completely soluble in cold nitric acid and in boiling sulphuric acid.

In the preparations containing metallic mercury the mercury is reduced to a fine state of subdivision with some inert substance. For this extinguishment various substances have been used, as chalk, honey, oleic acid, resinous tinctures, confection of rose, wool fat, mercury ointment, oleate of mercury. Mercury in its uncombined state is inert.

Official in ammoniac plaster with mercury, plaster, mercury with chalk, mass and ointment.

It is alterative. Dose: $\frac{1}{2}$ to 1 grain (0.03 to 0.065 gm.).

EMPLASTRUM AMMONIACI CUM HYDRARGYRO (Ammoniac Plaster with Mercury). Contains 18% of metallic mercury.

Prepared by emulsifying ammoniac with dilute acetic acid, evaporating to a syrupy consistence, adding the mercury which has been previously subdivided with oleate of mercury, and adding lead plaster to make the required quantity.

It is stimulant.

EMPLASTRUM HYDRARGYRI (Mercurial Plaster).

Contains 30% of metallic mercury.

Prepared by subdividing mercury with oleate of mercury and adding to lead plaster.

HYDRARGYRUM CUM CRETA (Mercury with Chalk), Grey Powder.

Contains 38% of metallic mercury, subdivided with prepared chalk and honey.

It is in smooth, greyish powder form, insoluble in water. It is a mild mercurial. Dose: 5 to 30 grains (0.33 to 1.95 gm.).

MASSA HYDRARGYRI (Mass of Mercury), Blue Mass, Blue Pill, Mercurial Pill.

Contains 33% of metallic mercury, made into a mass with licorice, marshmallow and honey of rose, with a small amount of glycerin to keep it in a soft condition.

It is a mild mercurial. Dose: 5 to 15 grains (0.33 to 0.97 gm.).

UNGUENTUM HYDRARGYRI (Mercurial Ointment), Blue Butter, Blue Ointment.

Contains 50% of metallic mercury, subdivided with oleate of mercury and mixed with lard and suet. It is easily absorbed by the skin.

UNGUENTUM HYDRARGYRI NITRATIS (Ointment of Mercuric Nitrate), Citrine Ointment.

Prepared by heating lard oil to 100°C (212°F) adding nitric acid, and when effervescence ceases, cooling to 40°C (104°F) and adding the mercury previously dissolved in the remainder of the nitric acid (heated to prevent crystallization) and setting aside to cool.

When freshly prepared it is a bright golden yellow color, but on exposure soon changes to a dirty greenish color. It should not be handled with an iron or steel spatula.

Used as a stimulant and alterative application.

HYDRARGYRI CHLORIDUM CORROSIVUM (Corrosive Chloride of Mercury) $HgCl_2$ Corrosive Sublimate, Bichloride of Mercury, Perchloride of Mercury, Mercuric Chloride.

Prepared by subliming together a mixture of persulphate of mercury, sodium chloride and manganese dioxide.

The persulphate of mercury is prepared by boiling mercury in an excess of sulphuric acid to dryness.

Permanent in the air. Soluble in water and alcohol. Its solubility in water is increased by the addition of a soluble chloride, as sodium or ammonium chloride. When added to alkalies the yellow oxide of mercury is precipitated. Yellow Wash (Lotio Hydrargyri Flava) is prepared by adding corrosive sublimate to lime water. When mixed with sulphurous or hypophosphorus acids or their salts it is reduced to calomel.

It produces precipitates in fusions or decoctions of chamomile, horseradish, columbo, catechu, cinchona, rhubarb, senna, simaruba, and oak bark. It is slowly changed into calomel by syrup sarsaparilla compound and syrup honey, but is not changed by pure syrup.

Antidotes: Whites of eggs, albumen in any form, which unites with the corrosive sublimate, forming insoluble compounds.

It is alterative and germacide. Dose: $\frac{1}{100}$ to $\frac{1}{8}$ gr.(0.0006 to 0.008 gm.).

Externally it is a stimulant, and escharotic.

Used in making ammoniated mercury.

HYDRARGYRUM AMMONIATUM (Ammoniated Mercury), White Precipitate, Mercuric Ammonium Chloride.

Prepared by adding ammonia water to a solution of corrosive sublimate, collecting, washing and drying.

Permanent in the air. Insoluble in water or alcohol. Often adulterated with white lead, chalk, calcium sulphate or starch.

Official in the ointment.

UNGUENTUM HYDRARGYRI AMMONIATI (Ointment of Ammoniated Mercury).

Contains 10% of ammoniated mercury, mixed with benzoinated lard.

HYDRARGYRI CHLORIDUM MITE (Mild Chloride of Mercury) Hg_2Cl_2 Mild Mercurous Chloride, Calomel, Protochloride of Mercury.

Prepared by subliming together a mixture of persulphate of mercury, mercury, and sodium chloride.

It should not be dispensed in conjunction with the chlorides of ammonium, sodium or potassium, hydrochloric acid, tincture of ferric chloride, or other soluble chlorides, owing to the liability of forming corrosive sublimate.

It may be distinguished from ammoniated mercury by heating with potassium or sodium hydrate test solution, and not giving an ammoniacal odor. From mercuric chloride by adding to lime water, mercuric chloride throws down yellow oxide of mercury, and mercurous chloride black oxide of mercury.

Black wash (Lotio Hydrargyri Nigra) is prepared by mixing together calomel and lime water.

Calomel sometimes contains corrosive sublimate as an impurity, which may be detected by boiling with water, filtering, adding silver nitrate test solution to the filtrate which, if corrosive sublimate be present will throw down a white precipitate of silver chloride. To remove corrosive sublimate from calomel, boil with water until the washings cease to give a precipitate with silver nitrate test solution.

Official in compound antimony pills and compound cathartic pills.

It is alterative, purgative and anthelmintic. Dose as an alterative: $\frac{1}{2}$ to 1 grain (0.03 to 0.065 gm.). As a purgative: 5 to 15 grains (0.33 to 1 gm.).

PILULÆ CATHARTICÆ COMPOSITÆ (Compound Cathartic Pills).

Contain compound extract of colocynth, calomel, extract of jalap and gamboge.

Each pill weighs about 3 grains.

HYDRARGYRI CYANIDUM (Mercuric Cyanide) $Hg(CN)_2$ Cyanuret of Mercury, Bicyanide of Mercury, Prussiate of Mercury.

Prepared by action on mercuric oxide with hydrocyanic acid, and should be kept in well-stoppered, dark amber bottles.

It becomes dark colored on exposure to light. Soluble in water and alcohol. It is a very poisonous salt.

It is alterative. Dose: $\frac{1}{16}$ to $\frac{1}{8}$ grain (0.004 to 0.008 gm.).

HYDRARGYRI IODIDUM FLAVUM (Yellow Mercurous Iodide) Hg_2I_2, Protiodide of Mercury, Green Iodide of Mercury, Iodide of Mercury.

Prepared by dissolving mercury in nitric acid, forming mercurous nitrate and acting on this with potassium iodide. It should be kept in dark, amber-colored bottles, away from the light.

On exposure to light it becomes darker, in proportion as it undergoes decomposition into metallic mercury and mercuric iodide. Almost insoluble in water and insoluble in alcohol.

It often contains as an impurity, red mercuric iodide, which may be detected by shaking the suspected salt with 10 C.c. of alcohol, filtering, and a portion of the filtrate tested with hydrogen sulphide should not give a precipitate.

Used principally in advanced syphilis. Dose: $\frac{1}{2}$ grain (0.033 gm.).

HYDRARGYRI IODIDUM RUBRUM (Red Mercuric Iodide) HgI_2, Biniodide of Mercury, Deutiodide of Mercury.

Prepared by acting on a solution of corrosive sublimate with one of potassium iodide, and should be kept in well-stoppered bottles, away from the light.

Permanent in the air. Almost insoluble in water and insoluble in alcohol. Used for similar purposes to that of mercurous iodide.

Used in making Donovan's solution. Dose: $\frac{1}{16}$ grain (0.004 gm.).

LIQUOR ARSENII ET HYDRARGYRI IODIDI (Solution of Arsenic and Mercuric Iodide), Solution of Hydriodate of Arsenic and Mercury, Donovan's Solution.

It is an aqueous solution containing 1% each of arsenic and mercuric iodides.

It should be a pale yellow color, and when found of an orange yellow it contains free iodine.

It is an alterative. Dose: 5 to 10 drops (0.3 to 0.6 C.c.).

HYDRARGYRI OXIDUM FLAVUM (Yellow Mercuric Oxide) HgO.

Prepared by pouring a solution of corrosive sublimate into one of caustic soda, and should be kept in well-stoppered bottles, away from the light.

Permanent in the air, but turns darker on exposure to the light. Insoluble in water and alcohol. Soluble in dilute hydrochloric and dilute nitric acids.

It differs from the red in being in a much finer state of subdivision.

It is distinguished from the red oxide by being precipitated by oxalic acid in the form of a white precipitate, while the red oxide is not.

Used in the local treatment of diseases of the eyes. Official in the ointment and oleate.

UNGUENTUM HYDRARGYRI OXIDI FLAVI (Ointment of Yellow Mercuric Oxide).

Contains 10% of yellow mercuric oxide mixed with simple ointment. It is used in scrofulous and syphilitic ulcers.

OLEATUM HYDRARGYRI (Oleate of Mercury).

Prepared by dissolving yellow mercuric oxide in oleic acid and contains 20%.

It is usually the consistence of petrolatum, owing to the formation of mercuric palmitate and stearate. It is more readily absorbed by the skin than the ointment.

HYDRARGYRI OXIDUM RUBRUM (Red Mercuric Oxide), HgO, Red Precipitate.

Prepared by dissolving mercury in nitric acid, heating and driving off the acid vapors.

Permanent in the air. Insoluble in water and alcohol. Soluble in nitric and hydrochloric acids. The two oxides of mercury have the same chemical composition, but differ in physical appearance.

Official in the ointment, and is used in making solution mercuric nitrate.

UNGUENTUM HYDRARGYRI OXIDI RUBRI (Ointment of Red Mercuric Oxide). Contains 10% of red mercuric oxide, 85% of simple ointment, and 5% of castor oil to subdivide the oxide.

LIQUOR HYDRARGYRI NITRATIS (Solution of Mercuric Nitrate).

Prepared by dissolving red mercuric oxide in nitric acid and water. Contains about 60% of mercuric nitrate and about 11% of free nitric acid, and should be kept in glass-stoppered bottles.

Used as a caustic application to ulcers.

HYDRARGYRI SUBSULPHAS FLAVUS (Yellow Mercuric Sulphate), $Hg(HgO)_2SO_4$, Basic Mercuric Sulphate, Turpeth Mineral.

Prepared by dissolving mercury in sulphuric acid and water, converting the mercurous to mercuric sulphate with nitric acid and heat, evaporating and powdering. It should be kept in well-stoppered bottles, away from the light.

Permanent in the air. Slightly soluble in water and Insoluble in alcohol. It is alterative, emetic and errhine, but is seldom used.

NITROGEN.

NITROGEN (N). The principal source of this element is the atmosphere, which consists of nearly four-fifths of nitrogen (the remaining fifth being almost entirely oxygen).

It may be prepared by burning phosphorus in a confined portion of air.

It is a colorless, tasteless and odorless gas.

Its chief use in the air is to dilute the energetic oxygen (a mechanical mixture resulting).

OXYGEN.

OXYGEN (O). It is the most abundant element in nature, although in a combined state, forming about one-half of the whole weight of our globe. The atmosphere consists of about one-fifth of its bulk of oxygen.

It may be prepared by heating potassium chlorate.

It is a colorless, tasteless and odorless gas, and is soluble in water to the extent of 3 volumes in 100 volumes, or fishes could not breathe.

There is a polymeric form of oxygen called OZONE, O_3, which may be prepared artificially by passing air through a box highly charged with electricity.

It is a powerful bleaching, disinfecting and oxidizing agent. Insoluble in water, soluble in oils of turpentine and cinnamon. It has a peculiar smell.

PHOSPHORUS COMPOUNDS.

PHOSPHORUS (Phosphorus), (P) is a non-metallic element obtained from bones by calcining, powdering, digesting with sulphuric acid previously diluted with water (the sulphuric acid separates a part of the lime from the calcium phosphate and precipitates it as calcium sulphate, while an acid calcium phosphate remains in solution) straining off the calcium sulphate, evaporating, again straining, evaporating to a syrupy consistence, mixing with powdered charcoal, and heating to a dull redness. The acid phosphate is changed into metaphosphate, which is distilled in a retort and the condensed phosphorus collected under water.

The bones of the sheep are the best for this purpose.

It is a semi transparent solid, without taste, but possessing an alliaceous smell. It is flexible, and when cut has a waxy lustre. When exposed to the air it emits white fumes, which are luminous in the dark and having an odor somewhat resembling garlic. On longer exposure to air it takes fire spontaneously. Insoluble in water and only partially soluble in alcohol. Soluble in ether and chloroform. On account of its great inflammability it should be kept under water.

Antidotes: Produce vomiting with copper sulphate given in dilute solution, three grains every five minutes, then give old turpentine, or better, a solution of potassium permanganate $\frac{1}{10}\%$.

Official in the pills, spirit and phosphorated oil. It is a general stimulant, especially for the brain. Dose: $\frac{1}{100}$ to $\frac{1}{75}$ of a grain (0.0006 to 0.0008 gm.).

There is another form of phosphorus called RED or AMORPHOUS PHOSPHORUS, made by exposing phosphorus to a temperature of 215°C (419°F) to 250°C (482°F) in an atmosphere which has no action upon it, or in closed glass tubes.

This variety is less easily acted upon by the air than ordinary phosphorus. Insoluble in carbon disulphide, ether or alcohol. It is used in the manufacture of matches and forms a much safer material than ordinary phosphorus.

SPIRITUS PHOSPHORI (Spirit of Phosphorus), Tincture of Phosphorus.

Prepared by dissolving phosphorus in boiling absolute alcohol. Each fluid drachm contains $\frac{1}{5}$ of a grain of phosphorus.

Official in the elixir.

ELIXIR PHOSPHORI (Elixir of Phosphorus).

Contains spirit of phosphorus, oil of anise, glycerin and aromatic elixir. Each fluid drachm contains about $\frac{1}{35}$ of a grain of phosphorus.

Dose: 20 to 40 minims (1.25 to 2.5 C.c.).

OLEUM PHOSPHORATUM (Phosphorated Oil).

Contains 1% of phosphorus.

Prepared by heating oil of sweet almond, straining, adding phosphorus and again heating, agitating until the phosphorus is dissolved and adding ether as a preservative and to render it less disagreeable to the taste.

The object in heating the oil is to expel air and traces of water which would aid in oxidizing the phosphorus.

Dose: 3 to 5 minims (0.18 to 0.3 C.c.).

PILULÆ PHOSPHORI (Pills of Phosphorus).

Prepared by dissolving phosphorus in chloroform with the aid of heat, mixing with powdered acacia and powdered marshmallow, previously mixed together, and make the mass with a mixture of two volumes of glycerin and one volume of water, then coat with balsam of tolu dissolved in ether. Each pill contains $\frac{1}{100}$ of a grain of phosphorus (0.0006 gm.).

POTASSIUM COMPOUNDS.

POTASSIUM (K) is a metal obtained by decomposing potassium carbonate, mixed with charcoal.

It has a strong affinity for oxygen, and when thrown upon water takes fire and burns with a rose-colored flame, combining with oxygen and generating potassium hydrate, which dissolves in the water.

On account of this property it must be kept in liquids, such as naptha, which contain no oxygen.

POTASSA (Potassa), KOH. Potassium Hydrate, Potassium Hydroxide, Caustic Potash.

Prepared by decomposing a solution of potassium carbonate with freshly slaked lime, filtering off the precipitated calcium carbonate, evaporating the filtrate to dryness, fusing and moulding into sticks or cakes.

It is very deliquescent and should be kept in hermetically sealed bottles. Soluble in $\frac{1}{2}$ part water and 2 parts of alcohol. It may be distinguished from the other fixed alkalies (soda and lithia) by its solution affording a crystalline precipitate (cream of tartar) with an excess of tartaric acid. It is a powerful escharotic.

Used in making the solution of potassa and potassa with lime.

LIQUOR POTASSÆ (Solution of Potassa). Contains 5% of potassa.

Prepared by dissolving 56 grams of potassa in 944 grams of water.

It should be kept in bottles with glass stoppers, which are coated with petrolatum, otherwise the stopper will be cemented to the neck of the bottle.

Incompatible with solutions of the metals, alkaloids, acid solutions, etc. Antidotes: Mild acids, as vinegar and lemon juice.

It is antacid and diuretic. Dose: 10 to 30 minims (0.6 to 1.9 C.c.).

POTASSA CUM CALCE (Potassa with Lime). Vienna Paste, Vienna Caustic.

Prepared by rubbing together equal parts of lime and potassa. Deliquescent and should be kept in well-stoppered bottles. It is used as a caustic, the lime causing it to be milder in its action.

POTASSA SULPHURATA (Sulphurated Potassa) Liver of Sulphur, Sulphuret of Potassium.

Prepared by heating together sublimed sulphur and potassium carbonate, and should be kept in well-stoppered bottles.

It is a mixture of potassium sulphide and hyposulphite, and is of a liver-brown color; therefore called liver of sulphur. Soluble in 2 parts of water.

On exposure to the air it attracts oxygen and the potassium sulphide is changed into potassium sulphate, which renders the preparation inodorous and white on the surface.

Its solution is decomposed by the mineral acids. It is incompatible with solutions of most of the metals, which are precipitated as sulphides.

It is a local irritant. Dose: 2 to 10 grains (0.13 to 0.65 gm.).

POTASSII ACETAS (Potassium Acetate) $KC_2H_3O_2$.

Prepared by neutralizing acetic acid with potassium carbonate and evaporating to dryness, liquifying by heat and cooling.

Deliquescent on exposure to air and should be kept in well-stoppered bottles.

Soluble in water and alcohol. It is used as a diuretic. Dose: 20 to 60 grains (1.3 to 3.9 gm.).

Incompatible with mineral acids, sodium and magnesium acetates, mercuric chloride and silver nitrate.

LECTURE NO. 9.

POTASSII BICARBONAS (Potassium Bicarbonate) $KHCO_3$. Saleratus.

Prepared by saturating a strong solution of potassium carbonate with carbonic acid gas, evaporating and crystallizing.

Permanent in the air. Soluble in water and almost insoluble in alcohol. Used in the preparation of solution of magnesium citrate and other effervescing draughts.

Used therapeutically for the same purposes as the carbonate. Dose: 20 to 60 grains (1.3 to 3.9 gm.).

POTASSII BICHROMAS (Potassium Bichromate) $K_2Cr_2O_7$, Potassium Dichromate, Red Chromate of Potash.

Prepared from chrome iron ore by heating with potassium nitrate and lixiviating the mass with water. The solution by evaporation yields potassium chromate, which is acted upon with sulphuric acid, forming the bichromate, potassium sulphate and water.

Permanent in the air. Soluble in water and insoluble in alcohol.

It is an irritant caustic, but is used mostly in volumetric solution and in the calico manufacture.

Antidotes: Soap, magnesia and chalk.

POTASSII BITARTRAS (Potassium Bitartrate) $KHC_4H_4O_6$, Cream of Tartar, Supertartrate of Potassa, Crystals of Tartar, Acid Potassium Tartrate.

Prepared from ARGOLS (which is an impure potassium bitartrate deposited on the sides of wine casks during fermentation) by treating with boiling water, the impurities filtered off and the solution evaporated and crystallized.

Permanent in the air. Slightly soluble in water, but its solubility is increased by the addition of borax or boracic acid. Sparingly soluble in alcohol. The U. S. P. allows 1% of impurities.

Often adulterated with gypsum, sand, chalk, etc.

Official in compound powder of jalap and compound effervescing powder.

Used as a cathartic and diuretic. Dose: 60 to 120 grains (3.9 to 7.8 gm.).

POTASSII BROMIDUM (Potassium Bromide). KBr.

Prepared by mixing bromine with a solution of potassa, evaporating to dryness, heating with charcoal, to convert the bromate present into bromide, dissolving the mass in water, filtering and crystallizing.

The bromate may be detected in the bromide by dropping on the crystals a drop of dilute sulphuric acid, which will produce a yellow color if bromate be present. It should be kept in well-stoppered bottles.

Permanent in the air. Soluble in water and but little soluble in alcohol.

Incompatible with solutions containing free chlorine, nitrous or nitric acids, bromine being freed. It is a sedative. Dose: 20 to 60 grains (1.3 to 3.9 gm.).

POTASSII CARBONAS (Potassium Carbonate) K_2CO_3, Sal Tartar.

Prepared from pearlash (impure potassium carbonate) by lixiviating with water, evaporating and granulating the product.

Very deliquescent. Soluble in water but insoluble in alcohol.

Incompatible with acids and acidulous salts, ammonium chloride, ammonium acetate, lime water, calcium chloride, magnesium sulphate, alum, tartar emetic, silver nitrate, ammoniated copper, ammoniated iron, ferrous sulphate, tincture ferric chloride, mercurous chloride, lead acetate, and subacetate, zinc sulphate. It is not decomposed by iron and potassium tartrate.

Used as an antacid and diuretic. Dose: 10 to 30 grains (0.65 to 1.95 gm.).

Impure potassium carbonate is called PEARLASH, or impure Potassa, and is obtained from the ashes of wood by lixiviation with water, filtering and evaporating.

POTASII CHLORAS (Potassium Chlorate) $KClO_3$. Chlorate of Potash, Hyperoxymuriate of Potassa.

Prepared by acting on a solution of potassium chloride, with calcium hypochlorite, with the aid of heat. Upon cooling the potassium chlorate crystallizes out and calcium chloride remains in solution. It should be kept in glass-stoppered bottles.

Permanent in the air. Soluble in water and insoluble in alcohol. It is slightly soluble in a mixture of water and alcohol. At 234°C (453.2° F) the salt fuses, and above 352°C (665.6°F) it is decomposed into oxygen and potassium perchlorate, and above 400°C (752°F) all its oxygen is liberated and potassium chloride remains.

It should not be mixed with organic matter, such as cork, tannic acid, sugar, etc., or with sulphur, antimony sulphide, phosphorus or other easily oxidizable substances, as dangerous explosions are apt to occur.

Used quite extensively as an alterative stimulant local application in inflammation of mucous membranes.

Dose internally: 10 to 20 grains (0.7 to 1.3 gm.). In overdose it is an active poison.

TROCHISCI POTASSII CHLORATIS (Troches of Potassium Chlorate). Contain potassium chlorate, sugar, tragacanth, spirit of lemon and water.

Each lozenge contains about 5 grains of the chlorate. Used mostly in sore throat.

POTASSII CITRAS (Potassium Citrate), $K_3C_6H_5O_7,H_2O$.

Prepared by adding potassium carbonate to a solution of citric acid in water, keeping the solution neutral, filtering and evaporating to dryness, and should be kept in well-stoppered bottles.

Deliquescent on exposure to air. Soluble in water and sparingly soluble in alcohol.

It is a grateful refrigerant diaphoretic. Dose: 20 to 25 grains (1.3 to 1.565 gm.).

POTASSII CITRAS EFFERVESCENS (Effervescent Potassium Citrate). Prepared by mixing in a warm mortar, citric acid, potassium bicarbonate and sugar in powder, drying and powdering.

It should be kept in well-stoppered bottles. Dose: 1 to 2 teaspoonfuls in cold water.

LIQUOR POTASSII CITRATIS (Solution of Potassium Citrate), Mixture of Citrate of Potash, Neutral Mixture, Saline Mixture, Effervescent Draught. Contains 9% of anhydrous potassium citrate.

Contains citric acid, potassium bicarbonate and water, and should be freshly prepared when wanted.

It is a refrigerant diaphoretic. Dose: ½ fluid ounce (15 C.c.).

POTASSII CYANIDUM (Potassium Cyanide), KCN. Cyanuret of Potassium.

Prepared by heating together potassium ferrocyanide and carbonate. The iron settles at the bottom, from which the fused liquid is poured into moulds and cooled.

Deliquescent in moist air, and should be kept in well-stoppered bottles. Soluble in water and sparingly soluble in alcohol. It should be white, and if yellow contains iron.

It is used largely in electro-metallurgy and photography. Internally it is used for the same purpose as dilute hydrocyanic acid. Dose: ⅛ grain (0.008 gm.).

It is very poisonous and the antidote is the same as that for hydrocyanic acid.

POTASSII ET SODII TARTRAS (Potassium and Sodium Tartrate) $KNaC_4H_4O_6, 4H_2O$. Tartrated Soda, Rochelle Salt.

Prepared by dissolving sodium carbonate in water, adding potassium bitartrate, boiling a few minutes, keeping the solution neutral, filtering, concentrating and crystallizing.

It slightly effloresces in dry air. Soluble in water and almost insoluble in alcohol. Often adulterated with sodium sulphate.

Incompatible with most acids, and with all acidulous salts except potassium bitartrate. Decomposed by lead acetate and subacetate, and by the soluble calcium and barium salts, unless the solution of the tartrate be considerably diluted.

It is a mild, cooling purgative. Dose: ½ to 1 ounce (15.5 to 31.1 gm.).

PULVIS EFFERVESCENS COMPOSITUS (Compound Effervescing Powder, Seidlitz Powder, Effervescing Tartrated Soda Powder.

Each powder contains 40 grains sodium bicarbonate mixed with 120 grains potassium and sodium tartrate folded in a blue paper, and 35 grains tartaric acid, folded in a white paper.

When taken the powders should be dissolved separately in water and then mixed.

Used as a laxative.

POTASSII FERROCYANIDUM (Potassium Ferrocyanide) $K_4Fe(CN)_6$, $3H_2O$ Yellow Prussiate of Potash, Ferrocyanuret of Potassium, Ferroprussiate of Potash, Prussiate of Potassa.

Prepared by heating animal matters, such as horns, hoofs, chips of horns, woolen rags, etc., with pearlash and scrap iron, in an iron pot, dissolving the mass in water, evaporating and crystallizing. The crystals contain some potassium cyanide which is converted into the ferrocyanide by the aid of ferrous sulphide and caustic potassa.

Slightly efflorescent on exposure to dry air, and should be kept in well-stoppered bottles. Soluble in water and insoluble in alcohol.

Used principally by dyers and calico printers, and for the preparation of dilute hydrocyanic acid. Rarely used internally. Dose: 10 to 15 grains (0.65 to 0.97 gm.).

POTASSII HYPOPHOSPHIS (Potassium Hypophosphite), KH_2PO_2.

Prepared by acting on a solution of calcium hypophosphite with one of granulated potassium carbonate, filtering off the precipitated calcium carbonate, evaporating and crystallizing.

Very deliquescent, and should be kept in well-stoppered bottles. Soluble in water and alcohol.

Official in syrup of hypophosphites. Used in the treatment of phthisis. Dose: 10 to 30 grains (0.65 to 1.95 gm.).

SYRUPUS HYPOPHOSPHITUM (Syrup of Hypophosphites).

Contains calcium hypophosphite, potassium hypophosphite, sodium hypophosphite, diluted hypophosphorus acid, sugar, spirit of lemon and water.

Each fluid drachm contains about $2\frac{1}{2}$ grains of the calcium salt, and not quite 1 grain each of the potassium and sodium salts.

Official in syrup of hypophosphites with iron.

Dose: 1 to 2 fluid drachms (3.7 to 7.5 C.c.).

SYRUPUS HYPOPHOSPHITUM CUM FERRO (Syrup of Hypophosphites with Iron).

Contains ferrous lactate, potassium citrate to retain the lactate in solution, and syrup of hypophosphites. It is chalybeate. Dose: 1 to 2 fluid drachms (3.7 to 7.5 C.c.).

POTASSII IODIDUM (Potassium Iodide), KI.

Prepared by mixing iodine with solution of potash, evaporating to dryness, fusing with charcoal, dissolving in water, filtering, evaporating and crystallizing. In the first part of the process iodide and iodate of potassium are formed, and by fusing with charcoal the iodate is converted into iodide.

To test for the presence of iodate in iodide, make an aqueous solution of potassium iodide, boil to expel all gases, add a little starch test solution and a few drops of dilute sulphuric acid, and if iodate is present a blue color will immediately appear.

Permanent in the air, and should be kept in well-stoppered bottles. Soluble in water and alcohol. It often contains potassium carbonate as an impurity.

Incompatible with alkaloids, calomel, mercurous and mercuric oxides, turpeth mineral, white precipitate, blue mass, metallic mercury, spirit of nitrous ether, etc. Its properties are the same as those of iodine.

Official in the ointment and compound solution of iodine.

Dose: 2 to 10 grains (0.13 to 0.65 gm.).

UNGUENTUM POTASSI IODIDI (Ointment of Potassium Iodide). Contains 12% of potassium iodide.

Contains potassium iodide, sodium hyposulphite to prevent the formation of free iodine, and benzoinated lard.

Used for scrofulous tumors, indolent ulcers, swellings, etc., and does not discolor the skin.

POTASSII NITRAS (Potassium Nitrate), KNO_3. Salt Petre.

It exists native in the earth in Chili, India, U. S., and different parts of Europe. Often associated with calcium nitrate. The saline earths which contain it are lixiviated with water, filtered through wood ashes (which converts the calcium nitrate into potassium nitrate), evaporated, filtered and set aside to crystallize. It is further purified by resolution and recrystallization.

Permanent in the air. Soluble in water and little soluble in alcohol. When fused and moulded, or formed into little cakes, it is called Crystal Mineral or SAL PRUNELLE.

Official in diluted nitrate of silver and potassium nitrate paper.

It is refrigerant, diuretic and diaphoretic. Locally it is irritant and stimulant.

CHARTA POTASSII NITRATIS (Potassium Nitrate Paper).

Prepared by dissolving potassium nitrate in water, dipping strips of white, unsized paper in this solution and drying. Used principally in asthma.

POTASSII PERMANGANAS (Potassium Permanganate), $KMnO_4$.

Prepared by evaporating to dryness a mixture of potassa, potassium chlorate, manganese dioxide with a little water, which forms potassium manganate. It is then heated to redness, cooled and boiled with water, which converts the manganate into permanganate.

It contains a large quantity of oxygen and is one of the most powerful oxidixing agents, causing the combustion of certain inflammable bodies, and imparts oxygen to almost all organic bodies.

It should be kept in well stoppered bottles, away from light. Permanent in the air. Soluble in water and is decomposed in contact with alcohol.

It is a powerful disinfectant. When used internally it may be made into pills with cacao butter or resin cerate. Dose : 1 to 2 grains (0.065 to 0.13 gm.).

POTASSII SULPHAS (Potassium Sulphate), K_2SO_4. Vitriolated Tartar.

Obtained as a secondary product in the preparation of nitric acid from potassium nitrate and sulphuric acid. Also prepared by the decomposition of potassium tartrate with calcium sulphate.

Permanent in the air. Soluble in water and insoluble in alcohol.

It is a mild purgative. Dose : 15 to 60 grains (0.975 to 3.90 gm.).

Silicon Compounds.

LIQUOR SODII SILICATIS (Solution of Sodium Silicate), Soluble Glass.

Prepared by fusing together sand and dried sodium carbonate, then boiling the mass with water. Contains about 20% of silica and 10% of soda.

Used in the preparation of mechanical dressings in surgery.

Silver Compounds.

ARGENTI CYANIDUM (Silver Cyanide), AgCN. Cyanuret of Silver, Cyanate of Silver.

Prepared by passing hydrocyanic acid gas into a solution of silver nitrate, or by mixing solutions of potassium cyanide and silver nitrate.

Permanent in dry air, but gradually turns brown when exposed to light, and should be kept in dark amber colored bottles, away from light. Insoluble in water or alcohol, but soluble in boiling nitric acid, ammonia water, solution of sodium hypophosphite, or of potassium cyanide. Its best solvent is potassium cyanide. Used for the extemporaneous preparation of dilute hydrocyanic acid.

ARGENTI IODIDUM (Silver Iodide), AgI.

Prepared by mixing solutions of potassium iodide and silver nitrate, washing and drying the precipitate. It should be kept in close stoppered bottles, away from the light. It is unalterable in the light if pure, but as usually found, becomes somewhat greenish yellow. Insoluble in water, alcohol, dilute acids, or in solution of ammonium carbonate. Soluble in 2500 parts of stronger ammonia water, and in an aqueous solution of potassium cyanide, or a concentrated solution of potassium iodide.

It has the general medicinal properties of silver nitrate.

Dose: 1 to 2 grains (0.065 to 0.13 gm.).

ARGENTI NITRAS (Silver Nitrate), $AgNO_3$.

Prepared by dissolving silver in nitric acid, evaporating and crystallizing.

It becomes gray, or grayish black, on exposure to light in the presence of organic matter. Soluble in water and alcohol. Its solution stains the skin, linen and muslin an indelible black color.

Incompatible with most all spring and river water, on account of a little common salt usually contained in them, with soluble chlorides, sulphuric, hydrosulphuric, hydrochloric, tartaric acids and their salts, with the alkalies and their carbonates, lime water and with astringent infusions.

Used in preparing moulded and fused nitrate of silver.

It is employed as an escharotic, and internally in the treatment of subacute gastritis, chronic diarrhœa, etc. Dose: $\frac{1}{4}$ grain (0.016 gm.) gradually increased to $\frac{1}{2}$ grain (0.032 gm.) made into a pill with sugar of milk and glucose. It should not be administered internally in solution as the solution is decomposed by the liquids of the mouth.

Antidote: Salt, soap, alkalies, etc.

ARGENTI NITRAS DILUTUS (Diluted Silver Nitrate), Silver and Potassium Nitrate, Mitigated Caustic.

Prepared by fusing together one part of silver nitrate with two parts of potassium nitrate, which renders the silver nitrate less caustic, and should be kept in dark amber colored bottles.

It becomes gray or grayish black on exposure to air in the presence of organic matter. It is escharotic.

ARGENTI NITRAS FUSUS (Fused Silver Nitrate) Lunar Caustic, Moulded Nitrate of Silver.

Prepared by fusing silver nitrate with 4% of hydrochloric acid, which renders the silver nitrate tough, and should be kept in dark amber colored bottles, away from the light.

It is liable to contain free silver, from having been exposed to too high a heat, lead and copper nitrates, from the impurity in the silver dissolved in the acid, and potassium nitrate, from fraudulent admixture, or otherwise.

ARGENTI OXIDUM (Silver Oxide), Ag_2O.

Prepared by precipitating a solution of silver nitrate with solution of lime.

It is reduced by the action of light, and should, therefore, be kept in well-closed bottles, away from the light. Slightly soluble in water and insoluble in alcohol.

Used sometimes as a substitute for nitrate.

Sodium Compounds.

SODIUM (Sodium) Na, is a soft, malleable, ductile solid of a silver white color.

When thrown upon water it acts the same as Potassium, forming Sodium Hydrate and the liberated hydrogen burns with a yellow flame. It should be kept under naptha.

SODA (Soda), NaOH. Sodium Hydrate, Sodium Hydroxide, Caustic Soda.

Prepared similar to potassa, by using sodium carbonate in place of potassium carbonate.

Its properties and uses are the same as those for potassa. Soluble in water and alcohol.

LIQUOR SODÆ (Solution of Soda.) Contains 5% of sodium hydrate.

May be prepared by dissolving 56 grams of soda in 944 grams of water. Its properties and uses are the same as those for solution of potassa.

SODII ACETAS (Sodium Acetate) $NaC_2H_3O_2, 3H_2O$.

Prepared by neutralizing acetic acid with sodium carbonate, evaporating and crystallizing.

Efflorescent in warm, dry air, and should be kept in well-stoppered bottles. Soluble in water and alcohol. It is diuretic. Dose: 20 to 120 grains (1.3 to 7.8 gm.).

SODII ARSENAS (Sodium Arsenate) $Na_2HAsO_4, 7H_2O$.

Prepared by fusing together arsenous acid, sodium nitrate, and dried sodium carbonate, dissolving the fused mass in boiling water, filtering and setting aside to crystallize.

Efflorescent in dry air, and should be kept in well-stoppered bottles. Soluble in water and slightly soluble in alcohol.

Its medical properties are the same as the other preparations of arsenic. Used in making the solution. Dose: $\frac{1}{12}$ to $\frac{1}{3}$ grains (0.005 to 0.02 gm.).

LIQUOR SODII ARSENATIS (Solution of Sodium Arsenate).

Prepared by dissolving sodium arsenate in distilled water, and contains 1%.

Dose: 3 to 5 minims (0.18 to 0.3 C.c.).

SODII BENZOAS (Sodium Benzoate) $NaC_7H_5O_2$.

Prepared by neutralizing a hot solution of sodium carbonate with benzoic acid, allowing to cool and crystallizing.

Permanent in the air. Soluble in water and alcohol.

Used as a remedy in gout and rheumatism. Dose: 1 to 2 drachms (3.9 to 7.8 gm.).

LECTURE NO. 10.

SODII BICARBONAS (Sodium Bicarbonate) $NaHCO_3$ Soda Saleratus.

Prepared by passing carbonic acid gas into a solution of sodium carbonate: also by reaction between sodium chloride and ammonium bicarbonate: also by the Solvay process as follows: By dissolving sodium chloride in ammonia water, passing a stream of carbonic acid gas into the solution, whereby sodium bicarbonate is precipitated and ammonium chloride remains in solution. It must contain 98.6% of the pure salt. Wrongly called Saleratus (which is potassium bicarbonate).

Permanent in the air, but slowly decomposes in moist air. Soluble in water and insoluble in alcohol.

Its medicinal properties are the same as those of the carbonate. Dose: 10 to 60 grains (0.65 to 3.9 gm.). Official in the troche.

TROCHISCI SODII BICARBONATIS (Troches of Sodium Bicarbonate). Contains sodium bicarbonate, sugar, nutmeg, and mucilage of tragacanth.

Each troche contains 3 grains of the salt. Dose: 1 to 6 troches.

SODII BISULPHIS (Sodium Bisulphite) $NaHSO_3$.

Prepared by thoroughly saturating a concentrated solution of sodium carbonate or bicarbonate with sulphurous acid gas, and collecting the crystals which form upon the cooling of the liquid. It should be kept in a cool place, in small, well-stoppered bottles, as nearly full as possible. When exposed to the air it loses sulphur dioxide and is gradually oxidized to sulphate. Soluble in water and partially so in alcohol.

When strongly heated it decrepitates, emits vapors of sulphur, and sulphur dioxide, and leaves a residue of sodium sulphate.

Soluble in water and sparingly so in alcohol.

SODII BORAS (Sodium Borate) $Na_2B_4O_7$, 10 H_2O. Borax, Biborate of Sodium, Pyroborate of Sodium.

Obtained in immense quantities native in the lakes of California, Thibet and Persia: and is found native in the earths of South America.

Prepared artificially by boiling together boric acid and sodium carbonate with water.

Slightly efflorescent in warm, dry air. Soluble in water and insoluble in alcohol. It has the property of rendering cream of tartar very soluble in water, and forms a combination called soluble cream of tartar. It is incompatible with the alkaloids, and in warm solution decomposes calomel.

Its medical properties are the same as those of boric acid. Dose: 30 to 40 grains (1.95 to 2.6 gm.).

SODII BROMIDUM (Sodium Bromide) NaBr.

Prepared the same as potassium bromide, solution of soda being used in place of solution of potassa, and should be kept in well-stoppered bottles.

On exposure to the air it absorbs water without deliquescing. Soluble in water and alcohol.

Its medicinal uses are the same as those of potassium bromide. Dose: ½ to 2 drachms (1.95 to 7.8 gm.).

SODII CARBONAS (Sodium Carbonate) $Na_2CO_3, 10H_2O$. Sal Soda. Washing Soda.

Obtained by three processes, namely: Le Blanc, Cryolite, and Solvay (or ammonia).

Le Blanc process: By acting on sodium chloride with sulphuric acid, then decomposing the sulphate with calcium carbonate.

Solvay process: By dissolving sodium chloride in ammonia water, and passing carbonic acid gas into the solution, then heating.

Cryolite process: Cryolite is a double salt of aluminum and sodium fluoride. Sodium carbonate is formed by heating cryolite with lime whereby calcium fluoride is formed, while the sodium and aluminum combine to form sodium aluminate, which is dissolved out by lixiviation. The soda is converted into the carbonate by passing carbonic acid gas under pressure into the solution, and the aluminum separated from the soda, becomes insoluble and is deposited.

Efflorescent in dry air. Soluble in water and insoluble in alcohol.

Incompatible with acids (which decompose it with effervescence), acidulous salts, lime water, ammonium chloride and earthy and metallic salts.

When taken in overdose it is a poison, the antidotes being fixed oils, acetic acid and lemon juice. Dose: 10 to 30 grains (0.65 to 1.95 gm.).

SODII CARBONAS EXSICCATUS (Dried Sodium Carbonate).

Prepared by breaking the crystals of sodium carbonate into small pieces, allowing to effloresce for several days, in warm air at a temperature not exceeding 20°C (77°F) and drying the white powder at a temperature of about 45°C (113°F) until the weight is one-half that of the quantity started with, passing through a sieve and keeping it in well-stoppered bottles.

Its advantage over the carbonate is that it can be readily made into pills.

Dose: 5 to 15 grains (0.33 to 1 gm.)

SODII CHLORAS (Sodium Chlorate) $NaClO_3$.

Prepared by adding a strong solution of tartaric acid to a hot aqueous solution of sodium carbonate, adding a solution of potassium chlorate. Potassium bitartrate will separate while the sodium chlorate in solution may be obtained by evaporating and crystallizing. It should be kept in glass-stoppered bottles, and the same precautions be used in handling it as are used in potassium chlorate.

Permanent in dry air. Soluble in water and glycerin, partially soluble in alcohol.

It has the same medicinal properties as potassium chlorate. Dose: 5 to 15 grains (0.3 to 1 gm.).

SODII CHLORIDUM (Sodium Chloride), NaCl, Table Salt, Common Salt, Sea Salt, Muriate of Soda.

It exists in nature in a solid state as Rock Salt and Fossil Salt in the earth, and in solution in certain mineral springs and the ocean.

Permanent in dry air when pure, but when contaminated with magnesium chloride, as it often is, it is deliquescent. Soluble in water and almost insoluble in alcohol.

Incompatible with some of the acids, (particularly sulphuric and nitric acids, which disengage vapors of hydrochloric acid) silver nitrate and mercurous nitrate.

It is used principally for domestic purposes as a condiment and antiseptic. In pharmacy it is used to prepare chlorine, hydrochloric acid, ammonium chloride, mild mercurous chloride, mercuric chloride, and sodium sulphate. It is tonic, anthelmintic and purgative. Dose: 2 drachms to one-half ounce (7.8 to 15.5 gm.).

SODII HYPOPHOSPHIS (Sodium Hypophosphite) NaH_2PO_2, H_2O.

Prepared by adding sodium carbonate to a solution of calcium hypophosphite as long as a precipitate of calcium carbonate is formed, filtering the solution and evaporating. In evaporating the solution care should be exercised and not too great a heat be employed, else explosion will result. It should be kept in well-stoppered bottles.

Deliquescent in moist air. Soluble in water and alcohol. Incompatible with the soluble salts of mercury and silver.

Its medical properties are the same as the other hypophosphites. Official in syrup of hypophosphites. Dose: 10 to 30 grains (0.65 to 1.95 gm.).

SODII HYPOSULPHIS (Sodium Hyposulphite) $Na_2S_2O_3$, $5H_2O$. Sodium Thiosulphite.

Prepared by heating a mixture of sodium carbonate with sulphur, forming sodium sulphide, stirring while hot so that every portion of it may come in contact with the air, which converts the sulphide to sulphite, dissolving in water, boiling with sulphur, which forms sodium hyposulphite, evaporating and crystallizing. It should be kept in well-stoppered bottles.

Permanent in the air, below 33°C (91.4°F) but efflorescent above that temperature. Soluble in water and insoluble in alcohol. Its solution dissolves silver chloride and all water insoluble compounds of silver (except the sulphide), iodine; it decomposes iodic acid with the liberation of iodine, and destroys the blue color of starch iodide.

It is used to prevent putrefaction and also used extensively in photography.

SODII IODIDUM (Sodium Iodide) NaI.

Prepared by the same process as potassium iodide except using solution of caustic soda in place of caustic potassa, and should be kept in well-stoppered bottles.

Deliquescent in moist air and becomes partially decomposed into sodium carbonate and free iodine. Soluble in water and alcohol.

Its medical properties are the same as those of potassium iodide. Dose: 20 grains (1.3 gm.).

SODII NITRAS (Sodium Nitrate) $NaNO_3$. Cubic Nitre.

It exists naturally in Chili, Peru and Brazil.

Deliquescent in moist air, and should be kept in well-stoppered bottles. Soluble in water and partially soluble in alcohol.

Used in making sulphuric acid and is rarely used internally. Dose: $\frac{1}{2}$ to 1 ounce (15.5 to 31.1 gm.).

SODII NITRIS (Sodium Nitrite) $NaNO_2$.

Prepared by heating sodium nitrate with organic substances or by fusing it with lead, and should be kept in well-stoppered bottles.

When exposed to the air it deliquesces and is gradually oxidized to sodium nitrate. Soluble in water and slightly soluble in alcohol.

It was made official in the U. S. P. 1890 solely for the preparation of spirit of nitrous ether.

SODII PHOSPHAS (Sodium Phosphate) $Na_2HPO_4, 12H_2O$, Sodium Orthophosphate, Medicinal Tribasic Phosphate of Sodium.

Prepared from bone ash or bone earth (obtained by burning bones with the access of air), by treating with sulphuric acid, forming acid phosphate of calcium, and saturating with a hot solution of sodium carbonate. It should be kept in well-stoppered bottles.

Effloresces in the air, and gradually loses 5 molecules of water of crystallization (25.1%). Soluble in water and insoluble in alcohol.

Incompatible with soluble salts of lime (with which it gives a precipitate of calcium phosphate), and the neutral metallic solutions.

It is purgative. Dose: 20 to 40 grains (1.3 to 2.6 gm.).

SODII PYROPHOSHAS (Sodium Pyrophosphate) $Na_4P_2O_7, 10H_2O$.

Prepared by heating sodium phosphate to redness, dissolving in water, filtering and crystallizing.

Permanent in cool air, but slightly efflorescent in dry air. Soluble in water and insoluble in alcohol. Its medicinal action is the same as that of the phosphate.

Used in preparing ferric pyrophosphate.

SODII SALICYLAS (Sodium Salicylate) $NaC_7H_5O_3$.

Prepared by the action of salicylic acid on sodium carbonate, in the presence of water, liberating carbon dioxide and leaving sodium salicylate in solution. It should be kept in well-stoppered bottles, away from light and heat.

Permanent in cool air. Soluble in water, alcohol and glycerin.

Incompatible with acids, salts of metals, solutions of many alkalies, antipyrin, etc.

Used for the same purposes as salicylic acid. Dose: 15 to 30 grains (1 to 2 gm.).

SODII SULPHAS (Sodium Sulphate) Na_2SO_4, $10H_2O$. Glauber Salt.

Obtained from the residue left in the manufacture of hydrochloric acid from sodium chloride, by neutralizing with sodium carbonate and crystallizing from solution in water. It exists in nature in many springs.

Effloresces rapidly in the air, and finally loses all of its water of crystallization. It should be kept in well-closed bottles. Soluble in water and glycerin, but insoluble in alcohol.

Incompatible with potassium carbonate, calcium chloride, salts of barium, lead acetate and subacetate, silver, if the solutions are strong.

It is a hydragogue cathartic. Dose: ½ to 1 ounce (15.5 to 31.1 gm.).

SODII SULPHIS (Sodium Sulphite) Na_2SO_3, $7H_2O$.

Prepared by the action of sulphurous acid on sodium carbonate or caustic soda, and should be kept in well-stoppered bottles, in a cool place.

Effloresces in the air, and is slowly oxidized to the sulphate. Soluble in water, but insoluble in alcohol.

It is antiseptic. Dose: 1 drachm (3.9 gm.).

SODII SULPHOCARBOLAS (Sodium Sulphocarbolate) $NaSO_3C_6H_4(OH).2H_2O$.

Prepared by dissolving phenol in excess of sulphuric acid, supersaturating the liquid with barium carbonate, filtering, and treating the filtrate with sodium carbonate until no further precipitation occurs, filtering and evaporating.

Efflorescent in dry air. Soluble in water and less so in alcohol.

It unites the properties of sulphuric and carbolic acids. Dose: 3 to 20 grains (0.2 to 1.3 gm.).

Strontium Compounds

STRONTII BROMIDUM (Strontium Bromide) $SrBr_2$, $6H_2O$.

Prepared by burning strontium in bromine vapor. Also by dissolving strontium carbonate in hydrobromic acid. Very deliquescent. Soluble in water and alcohol, and should be kept in glass-stoppered bottles.

Incompatible with soluble carbonates and sulphates, all the other bromides and mineral acids.

Used for the same purpose as potassium bromide. Dose: 5 to 10 grains (0.33 to 0.66 gm.).

STRONTII IODIDUM (Strontium Iodide) $SrI_2, 6H_2O$.

Prepared by saturating a solution of hydriodic acid with strontium hydrate and evaporating. It should be kept in dark amber glass-stoppered bottles.

Very deliquescent and is colored yellow by exposure to air and light. Soluble in water but insoluble in alcohol. It contains 56.5% iodine.

Used for the same purpose as potassium iodide. Dose: 5 to 10 grains (0.33 to 0.66 gm.).

STRONTII LACTAS (Strontium Lactate) $Sr (C_3H_5O_3)_2, 3H_2O$.

Prepared by dissolving freshly precipitated strontium carbonate in lactic acid, evaporating and granulating.

Permanent in the air. Soluble in water and alcohol.

It is diuretic. Dose: 10 to 30 grains (0.65 to 1.95 gm.).

Sulphur Compounds.

SULPHUR SUBLIMATUM (Sublimed Sulphur) (S.) Flowers of Sulphur, Brimstone.

It exists native and also in combination with other metals, usually iron as iron pyrites, and is obtained by separating by means of heat and subliming.

It often contains traces of arsenic and sulphuric acid. The arsenic may be detected by washing the sulphur with ammonia water, acidulating with hydrochloric acid and precipitating with hydrogen sulphide.

Permanent in the air. Insoluble in water, but soluble in alkaline solution, petrolatum, rectified coal naptha, fixed oils, oil of turpentine and other volatile oils, alcohol, ether, chloroform, and carbon disulphide.

It is laxative and diaphoretic. Dose: 1 to 3 drachms (3.9 to 11.65 gm.). Used in making washed and precipitated sulphur. Roll Sulphur is crude sulphur that has been melted, poured into moulds and cooled.

SULPHUR LOTUM (Washed Sulphur).

Prepared by washing sulphur with ammonia water and water to remove any sulphuric acid and arsenic that may be present.

Official in the ointment and compound licorice powder.

Dose: ½ drachm to ½ ounce (2. to 15.5 gm.).

UNGUENTUM SUPHURIS (Ointment of Sulphur).

Prepared by rubbing 300 grams of washed sulphur with 700 grams of benzoinated lard. It is a specific for the itch.

SULPHUR PRECIPITATUM (Precipitated Sulphur) Milk of Sulphur.

Prepared by boiling sulphur with lime (forming sulphide of lime) then precipitating with hydrochloric acid, previously diluted with water.

It should be kept in well-stoppered bottles.

It differs from the sublimed sulphur in being in a finer state of subdivision, and in presenting, after fusion, a softer and less brittle mass.

Its medical properties are the same as those of sublimed sulphur.

SULPHURIS IODIDUM (Sulphur Iodide).

Prepared by fusing together washed sulphur and iodine (20 grains of sulphur and 80 grains of iodine). It should be kept in well-stoppered bottles, in a cool place.

Insoluble in water, but soluble in carbon disulphide and alcohol; ether dissolves out the iodine, leaving the sulphur. On exposure to the air it gradually loses iodine. Used as an external remedy in skin diseases.

Zinc Compounds.

ZINCUM (Zinc) Zn. is a metallic element in the form of thin sheets, also in irregular granulated pieces or moulded into thin pencils, or in a state of fine powder.

Obtained native as a sulphide called Blende, and as a carbonate and silicate, called Calamine.

LIQUOR ZINCI CHLORIDI (Solution of Zinc Chloride). Contains about 50% by weight of the salt.

Prepared by dissolving zinc in hydrochloric acid and water, adding nitric acid, evaporating to dryness dissolving in water, adding precipitated zinc carbonate, setting aside to separate the precipitate, and decanting the clear liquid. It is a disinfectant.

Burnett's Disinfecting Fluid is an aqueous solution of zinc chloride containing 200 grains in each Imperial fluid ounce.

ZINCI ACETAS (Zinc Acetate) Zn $(C_2H_3O_2)_2$, $2H_2O$.

Prepared by digesting zinc oxide in acetic acid and water, filtering, evaporating and crystallizing. It should be kept in well-stoppered bottles.

Exposed to the air it gradually loses some of its acid and effloresces. Soluble in water and alcohol. It is decomposed by the mineral acids, with the escape of acetous vapors.

It is astringent and is used locally.

ZINCI BROMIDUM (Zinc Bromide) $ZnBr_2$.

Prepared by the direct combination of zinc with bromine, or by dissolving zinc in solution of hydrobromic acid and should be kept in small, glass stoppered bottles. Deliquescent on exposure to the air. Soluble in water and alcohol.

Used in epilepsy. Dose: 1 to 2 grains (0.065 to 0.13 gm.).

ZINCI CARBONAS PRECIPITATUS (Precipitated Zinc Carbonate) $(ZnCO_3)_2Zn(OH)_2$.

Prepared by boiling together a solution of sodium carbonate and one of zinc sulphate, collecting the precipitate, draining and drying.

Permanent in the air. Insoluble in water or alcohol, but soluble in dilute acids and ammonia water.

Used externally.

ZINCI CHLORIDUM (Zinc Chloride) $ZnCl_2$.

Prepared by dissolving zinc in hydrochloric acid and water, boiling and making up the loss by evaporation with water, cooling with frequent stirring, adding a solution of chlorine (which combines with any iron present to form ferric chloride) adding zinc carbonate (the zinc combines with the chlorine to increase the product of zinc chloride, ferric oxide is precipitated, and carbonic acid is set free), filtering and evaporating. It should be kept in small, glass-stoppered bottles.

Very deliquescent on exposure to the air. Soluble in water and alcohol. It sometimes contains some oxychloride which remains undissolved in water. The commercial chloride sometimes contains as high as 12% of zinc arsenate as an impurity.

Internally it is alterative and antispasmodic, but is little used, except as a disinfectant. Antidotes: Alkalies and their carbonates or soap.

ZINCI IODIDUM (Zinc Iodide) ZnI_2.

Prepared by digesting zinc and iodine in the presence of water, or by saturating hydriodic acid with the oxide or carbonate of zinc.

Very deliquescent and is liable to absorb oxygen from the air, and to become brown from the liberated iodine.

It is alterative. Dose: ½ to 2 grains (0.03 to 0.13 gm.).

ZINCI OXIDUM (Zinc Oxide) ZnO.

Prepared by calcining carbonate of zinc until a portion ceases to effervesce when added to acids and should be kept in well-stoppered bottles.

The commercial oxide is very impure, usually containing the carbonate, subsulphate and often oxychloride.

When exposed to the air it gradually absorbs carbon dioxide.

Insoluble in water or alcohol, but soluble in dilute acids or ammonia water.

It is antispasmodic and astringent, but is employed principally as an application to excoriated surfaces in the form of an ointment.

Official in the ointment and oleate. Dose: 2 to 8 grains (0.13 to 0.52 gm.).

UNGUENTUM ZINCI OXIDI (Ointment of Zinc Oxide) Zinc Ointment.

Contains 20% of zinc oxide mixed with benzoinated lard.

Used as an astringent application.

OLEATUM ZINCI (Oleate of Zinc).

Prepared by dissolving 5% of zinc oxide in oleic acid. Used externally and was introduced for the first time into the U. S. P., 1890.

LECTURE NO. 11.

ZINCI PHOSPHIDUM (Zinc Phosphide) Zn_3P_2.

Obtained by passing vapors of phosphorus in a current of hydrogen over fused zinc.

It should be kept away from the air, in small glass-stoppered vials, to prevent the oxidation of the phosphorus. In contact with air it slowly emits phosphorus vapors. It should not be combined with vegetable extracts, since the presence of any vegetable acid will cause decomposition. Insoluble in water or alcohol, but soluble in dilute hydrochloric or sulphuric acids.

It acts about the same as phosphorus therapeutically. Dose: $\frac{1}{20}$ of a grain (0.003 gm.)

ZINCI SULPHAS (Zinc Sulphate) $ZnSO_4, 7H_2O$, White Vitriol.

Prepared by the action of dilute sulphuric acid on metallic zinc, hydrogen gas being evolved.

Efflorescent in dry air. Soluble in water and glycerin, but insoluble in alcohol.

Incompatible with alkalies and alkaline carbonates, sulphides, lime water, soluble lead salts, and astringent infusions. Antidotes. Bland drinks, opium, etc.

It is tonic, astringent and emetic. Dose: $\frac{1}{4}$ to $\frac{1}{2}$ grain (0.018 to 0.037 gm.), and as an emetic: 10 to 30 grains (0.65 to 1.95 gm.).

ZINCI VALERIANAS (Zinc Valerianate) $Zn(C_5H_9O_2)_2, 2H_2O$.

Prepared by interaction between hot solutions of zinc sulphate and sodium valerianate.

On exposure to air it slowly loses valerianic acid. Soluble in 100 parts of water and 40 of alcohol.

It is antispasmodic. Dose: 1 to 2 grains (0.05 to 0.13 gm.).

VOLUMETRIC ANALYSIS.

It should be the duty of every pharmacist to test his preparations to ascertain whether they are of the required strength or not, and the simplest way in which this can be done is by Volumetric Analysis.

An illustration is necessary, to explain the general principles of chemical action, upon which are based the methods of this form of analysis.

The reactions which take place between elements or compounds are expressed by means of equations, thus: $NaOH + HCl = NaCl + H_2O$.

This equation shows the reaction that takes place between sodium hydrate and hydrochloric acid, forming sodium chloride and water, and also the exact proportions by weight, in which the alkali and acid unite, and the quantities of the resulting products by weight, thus:

The molecular weight of bodies is found by adding together the atomic weights of the elements entering into the molecule, thus:

The atomic weight of Na is 23
" " " " O is 16
" " " " H is 1
———
40 the molecular weight of sodium hydrate.

The atomic weight of H is 1.
" " " " Cl is 35.4
———
36.4 the molecular weight of hydrochloric acid.

The atomic weight of Na is 23.
" " " " Cl is 35.4
———
58.4 the molecular weight of sodium chloride.

The atomic weight of H is 1.×2, equals 2.
" " " " O is 16.
———
18. the molecular weight of water.

It may be seen that 40 parts of sodium hydrate will be decomposed by 36.4 parts of hydrochloric acid, producing 58.4 parts of sodium chloride and 18 parts of water (the parts are by weight).

Matter is not destroyed, hence the sum of the weights on each side of the equation must be the same, thus:

$NaOH + HCl = NaCl + H_2O$
40   36.4   58.4   18
40+36.4=76.4   58.4+18=76.4

Sodium hydrate and hydrochloric acid will only unite in the above proportion, and if different quantities of the substances from those given be used, reaction will take place in that proportion, and the excess of either will remain uncombined.

Volumetric Analysis consists in noting the amount of a solution of known strength which is to be added to the substance in solution to produce a certain reaction, and upon the quantity required is based the calculation for the weight of the substance being analyzed.

When a sufficient quantity of the solution has been added, is ascertained by known changes, which in some cases may be seen by the eye, and in others when it is not visible to the eye, a substance is added.

,called an INDICATOR, which shows the end of the reaction by a change of color.

The following changes show that the end of the reaction has taken place :

1–When a precipitate is produced, as in the estimation of cyanogen by silver nitrate.

2–When a precipitate ceases to be formed, upon addition of more volumetric solution, as in the estimation of the chlorides by silver nitrate.

3–A change in color of the solution in which indicators are used, as in the estimation of acids and alkalies.

The principal indicators are :

Brazil wood, which turns purplish red with alkalies, and yellow with acids.

Cochineal, which turns violet with alkalies, and yellowish red with acids.

Litmus : Blue litmus turns red with acids and red litmus turns blue with alkalies.

Methyl Orange, which turns yellow with alkalies, hydrates, carbonates, bicarbonates. Carbonic acid does not affect it, but sulphuric hydrochloric and other acids change its color to crimson. It is not suited for use with organic acids.

Phenolphtalein, which turns deep purplish red with alkali hydrates and carbonates. Bicarbonates and most other salts do not produce such color. Acids render the reddened solution colorless. It should not be used as an indicator for ammonia or bicarbonates.

Rosolic Acid, which turns violet red with alkalies, and yellow with acids.

Tumeric, which turns brown with alkalies, and the yellow color is restored by acids. Boric acid, even in the presence of hydrochloric acid turns the color to reddish brown, which is changed to bluish black by ammonia.

Volumetric Solutions may be prepared with but little difficulty by the pharmacist.

A NORMAL VOLUMETRIC SOLUTION, designated $\frac{n}{1}$ is one that contains in one liter, the molecular weight of the reagent expressed in grams, and reduced to the valency corresponding to one atom of hydrogen or its equivalent.

Ex. Hydrochloric acid, molecular weight 36.4, having but one atom of basic hydrogen, has 36.4 grams of hydrochloric acid in one liter of the normal volumetric solution.

Sulphuric acid (H_2SO_4), molecular weight 97.82, having two atoms of replaceable hydrogen, contains one-half, or 48.91 grams of sulphuric acid in one liter of normal volumetric solution.

A DECINORMAL VOLUMETRIC SOLUTION, designated $\frac{N}{10}$ contains one-tenth of the quantity of the reagent used in a normal solution.

A CENTINORMAL VOLUMETRIC SOLUTION, designated $\frac{N}{100}$ contains one one-hundredth the amount.

A SEMINORMAL VOLUMETRIC SOLUTION, designated $\frac{N}{2}$ contains one-half the amount.

A DOUBLE NORMAL VOLUMETRIC SOLUTION, designated $\frac{2}{N}$ contains twice the amount.

The NORMAL FACTOR is the amount by which the number of cubic centimeters of normal solution used is to be multiplied to obtain the quantity of pure substance in the sample examined.

Ex. As before stated the molecular weight of sodium hydrate is 40, and 40 parts require 36.4 parts by weight of hydrochloric acid to decompose it. Sodium being a univalent element, each liter of the normal volumetric solution contains 40 grams of sodium hydrate.

If 1 liter (1000 C c.) contains 40 grams, 1 C.c. will contain .04 gm.

If 40 grams of sodium hydrate require 36.4 grams of hydrochloric acid to decompose it, .04 gm. would require .0364 of hydrochloric acid, therefore the normal factor of sodium hydrate is .04 and hydrochloric acid .0364.

The method of performing Volumetric Analysis is as follows, using the estimation of hydrochloric acid as an example :

Place in a beaker 2 gms. of hydrochloric acid, and add about 5 C.c. of water and a few drops of phenolphtalein solution, as indicator.

Place the beaker beneath the burette, and run in from the burette the normal sodium hydrate solution, little by little, until the pink tint produced is permanent ; this indicates the end of the reaction.

If 17.5 C.c. of the sodium hydrate solution are required, we multiply 17.5 by the normal factor of hydrochloric acid .0364 ; 17.5 + .0364 = .637 gm. therefore 2 gms. of hydrochloric acid contain .637 gm. of absolute acid, or 31.9%.

A detailed account of making the different volumetric solutions, indicators and the normal factors, etc., may be found in the U. S. Dispensatory, commencing on page 1776.

Before commencing Organic Pharmacy, I wish to have the manner of uniting elements to form compounds, and of balancing equations, thoroughly understood.

To do this, it is necessary to learn the LAWS REGULATING CHEMICAL COMBINATION (by volume or weight).

1st Law : A definite compound always contains the same elements and the same proportion of those elements (by weight or volume).

Ex. Hydrochloric acid contains 1 part of hydrogen and 1 part of chlorine by volume: The atomic weight of hydrogen is 1, and that of chlorine is 35.4, the molecule of hydrochloric acid would therefore contain 1 part of hydrogen and 35.4 of chlorine by weight. If hydrogen and chlorine were brought together in any other proportion, they would unite in the above proportion, and the excess of either element would remain uncombined.

2nd Law: Where two elements unite in more than one proportion they do so in simple multiples of that proportion.

This is why some elements have more than one valence.

Ex. The compounds of nitrogen and oxygen, thus:

Nitrogen Monoxide N_2O.
Nitrogen Dioxide N_2O_2.
Nitrogen Trioxide N_2O_3.
Nitrogen Tetroxide N_2O_4.
Nitrogen Pentoxide N_2O_5.

Nitrogen unites with oxygen in these five proportions.

(1)–2 volumes of nitrogen (28 parts by weight) and 1 volume of oxygen (16 parts by weight).

(2)–2 volumes of nitrogen (28 parts by weight) and 2 volumes of oxygen (32 parts by weight).

(3)–2 volumes of nitrogen (28 parts by weight) and 3 volumes of oxygen (48 parts by weight).

(4)–2 volumes of nitrogen (28 parts by weight) and 4 volumes of oxygen (64 parts by weight).

(5)–2 volumes of nitrogen (28 parts by weight) and 5 volumes of oxygen (80 parts by weight).

The first, third and fifth show the oxygen valence of nitrogen (N^i, N^{iii}, N^v,) and unite with water to form the corresponding acids: while the second and fourth are mixed oxides and do not form acids with water.

It may be seen from these compounds that the second and the others have exactly 2, 3, 4, and 5 times as much oxygen as the first, the quantity of nitrogen remaining the same.

3rd Law: The proportions in which two elements unite with a third are the proportions in which they unite with each other.

Ex. The atomic weight of oxygen is 16 and that of carbon 12. Carbon in proportions of 12 unites with hydrogen, and oxygen in proportions of 16 unites with hydrogen, therefore 12 and 16 or a multiple of 16 are the proportions in which carbon and oxygen will unite with each other.

Carbon Monoxide $CO(C_{12}O_{16})$.
Carbon Dioxide $CO_2(C_{12}O_{32})$.

To be able to balance equations, the valence of the principal elements should be committed to memory as follows:

		ATOMIC WEIGHT.			ATOMIC WEIGHT.
Aluminum	$(Al_2)^{vi}$	27.04	Lead	$Pb^{ii, iv}$	206.40
Ammonium	NH_4^i	18.01	Lithium	Li^i	7.01
Antimony	$Sb^{iii, v}$	119.60	Magnesium	Mg^{ii}	24.30
Arsenic	$As^{iii, v}$	74.90	Manganese	$Mn^{ii, iv, vi}\ (Mn_2)^{vi}$	54.80
Barium	$Ba^{ii, iv}$	136.90	Mercury	$Hg^{ii}\ (Hg_2)^{ii}$	199.80
Bismuth	$Bi^{iii, v}$	208.90	Nitrogen	$N^{i, iii, v}$	14.01
Bromine	$Br^{i, iii, v, vii}$	79.76	Oxygen	O^{ii}	15.96
Calcium	$Ca^{ii, iv}$	39.91	Phosphorus	$P^{iii, v}$	30.96
Carbon	C^{iv}	11.97	Potassium	K^i	39.03
Chlorine	$Cl^{i, iii, v, vii}$	35.37	Silver	Ag^i	107.66
Copper	$Cu^{ii}\ (Cu_2)^{ii}$	63.80	Sodium	Na^i	23.00
Gold	$Au^{i, iii}$	196.70	Strontium	$Sr^{ii, iv}$	87.30
Hydrogen	H^i	1.	Sulphur	$S^{ii, iv, vi}$	31.98
Iodine	$I^{i, iii, v, vii}$	126.53	Zinc	Zn^{ii}	65.10
Iron	$Fe^{ii}\ (Fe_2)^{vi}$	55.88			

Under chemical terms the definitions were given for Monads, Diads and Triads, but for convenience I will repeat them here.

Elements having one bond are termed MONADS, and are called Univalent: those having two bonds, DIADS (Bivalent): those having three bonds TRIADS (Trivalent): those having four bonds, TETRADS (Tetrivalent): those having five bonds, PENTADS (Pentivalent): those having six bonds, HEXADS (Hexivalent).

The elements with which we come in contact in pharmacy are mostly monads, diads and triads.

One monad will unite with, or replace one Monad; two monads will unite with, or replace a Diad; three monads will unite with, or replace a Triad, and so on. If we wish to unite a diad with a triad we would use two of the triad and three of the diad, thus: A diad having two bonds multiplied by three would have the same value as a triad having three bonds multiplied by two.

Elements having an even number of bonds, as Diads, Tetrads and Hexads, are called ARTIADS. Those having an uneven number of bonds, as Monads, Triads and Pentads, are called PERISSADS.

The acid radicals have the same valence as the number of atoms of basic hydrogen they contain. Ex., HCl: here we have one atom of basic hydrogen, therefore Cl is a monad radical: H_2SO_4, we have two atoms of basic hydrogen, therefore SO_4 is a diad radical: H_3PO_4, we have three atoms of basic hydrogen, therefore PO_4 is a triad radical.

By committing to memory the valences of the elements and the formulas for the acids, the formulas of the various pharmacopœial salts may be easily made. To form Ferrous Sulphate : Iron (Fe) in the ferrous compounds is a diad. The sulphuric acid radical (SO_4) is also a diad, then 1 atom of Fe will unite with one molecule of SO_4, thus : $FeSO_4$.

Ferric Sulphate : Iron in the ferric compounds is a triad, but 2 atoms of iron (2 triads) replace 6 atoms of hydrogen. As the SO_4 radical is a diad it would require 3 molecules of it (3 diads) to unite with the 2 atoms of iron, thus : $Fe_2(SO_4)_3$.

Ferrous Chloride : Iron being a diad and chlorine a monad, there will be required 2 atoms of Cl to unite with 1 atom of Fe, thus : $FeCl_2$.

Ferric Chloride : If 2 atoms of iron replace 6 atoms of hydrogen in the ferric compounds, there would be required 6 atoms of Cl (6 monads) to unite with the 2 atoms of Fe. Thus : Fe_2Cl_6.

In balancing equations it is necessary to know what bodies are formed by the resulting chemical action, then apply the valences thus : Write the symbols or formulas for the elements or compounds to be used with a plus sign between and followed to the right by an equals sign ; then to the right of the equals sign, write the bodies produced by the chemical change.

Ex. To make Zinc Sulphate from zinc and sulphuric acid.

$Zn + H_2SO_4 = X$.

Zn is a diad, SO_4 is a diad, then the Zn will unite with the SO_4 thus : $Zn + H_2SO_4 = ZnSO_4 + X$.

The hydrogen of the acid is set free, thus :

$Zn + H_2SO_4 = ZnSO_4 + 2H$.

The number of bonds to the left of the equals sign must be the same as those to the right.

To make Mercuric Iodide, from mercuric chloride and potassium iodide.

$HgCl_2 + KI = X$.

Hg in the mercuric compounds is a diad and I is a monad, then we must use 2 of I to unite with 1 of mercury. If we increase the amount of an element in a compound we must increase all the elements in the compound in the same proportion, according to our first law of chemical combination hence, we must use 2 molecules of KI. thus :

$HgCl_2 + 2KI = X$.

If Hg is a diad and 2I (2 monads) has the value of a diad, the 1 atom of Hg will unite with the 2 atoms of I thus :

$HgCl_2 + 2KI = HgI_2 + X$.

We have left, 2 monads (2Cl) and 2 monads (2K) which will unite with each other thus :

$HgCl_2 + 2KI = HgI_2 + 2KCl$.

To make Potassium Acetate from potassium carbonate and acetic acid.
$K_2CO_3 + HC_2H_3O_2 = X.$

In potassium carbonate we have 2 monads (2K) and in acetic acid we have 1 monad radical, therefore we must have 2 molecules of monad radical to unite with the 2 atoms of potassium (K) thus:

$K_2CO_3 + 2HC_2H_3O_2 = X.$

The two atoms of K will unite with the 2 molecules of $C_2H_3O_2$ thus:
$K_2CO_3 + 2HC_2H_3O_2 = 2KC_2H_3O_2 + X.$

When carbonates are acted on by an acid, carbonic acid gas (CO_2) is given off. In this reaction the carbonic acid radical splits up into carbon dioxide and oxygen, thus:

$K_2CO_3 + 2HC_2H_3O_2 = 2KC_2H_3O_2 + CO_2 + X.$

The oxygen being a diad unites with the two atoms of H (2 monads) from the acetic acid, thus:

$K_2CO_3 + 2HC_2H_3O_2 = 2KC_2H_3O_2 + CO_2 + H_2O.$

By continuous practice in writing and balancing equations you will find it of great value in many operations performed by the pharmacist.

Organic Pharmacy.

Acids.

ACIDUM ACETICUM (Acetic Acid) $HC_2H_3O_2$.

An organic acid containing 36% by weight of absolute acid and 64% by weight of water.

Obtained by the destructive distillation of wood, preferably oak; also by the oxidation of alcohol (called the quick vinegar process) which consists in allowing a weak alcoholic liquid to drip slowly over clean wood shaving in a cask, with a free access of air. The alcohol is first changed to aldehyde, which is then oxidized to acetic acid.

Wood when charred yields many volatile products, among which are an acid layer or liquor, an empyreumatic oil, and tar containing creosote and other proximate principles. The acid portion of the distillate is called PYROLIGNEOUS ACID or impure acetic acid from which the acetic acid is obtained.

It mixes with alcohol and water in all proportions. It is refrigerant and astringent, but is rarely used internally.

ACIDUM ACETICUM DILUTUM (Diluted Acetic Acid).

Contains 6% by weight of absolute acid, and is prepared by mixing together 1 part by weight of acetic acid and 5 parts by weight of water.

It is used as a substitute for vinegar because distilled vinegar contains a little organic matter, which is always darkened or precipitated when its acid is saturated with an alkali while dilute acetic acid does not produce this change.

LECTURE NO. 12.

ACIDUM ACETICUM GLACIALE (Glaciale Acetic Acid). Radical Vinegar.

Contains nearly 99% by weight of absolute acid.

Prepared by heating sodium acetate to drive off the water, and distilling with sulphuric acid. It should be kept in well-stoppered bottles.

It is a valuable solvent for the active constituent of drugs containing volatile oils, fixed oils, and dissolves camphor, gum-resins, etc.

Used externally as a rubifacient, vesicant and caustic.

ACIDUM BENZOICUM (Benzoic Acid) $HC_7H_5O_2$.

An organic acid obtained from benzoin by sublimation or artificially from toluol (a coal tar product). The acid obtained by the sublimation of benzoin is usually characterized by a fragrant odor : that obtained from toluol usually contains chlorine. It should be kept in dark amber, well-stoppered bottles, in a cool place.

It is somewhat volatile at a moderately warm temperature, and is rendered darker by exposure to the light. Sparingly soluble in water, but soluble in alcohol, ether and chloroform. Its solubility is increased in water by the addition of borax or sodium phosphate. The acid obtained from benzoin has a lower melting point and a greater solubility in water.

There is a German Benzoic Acid which is prepared from the urine of horses and cattle.

It is expectorant. Dose : 10 to 30 grains (0.65 to 1.9 gm.).

ACIDUM CARBOLICUM (Carbolic Acid) C_6H_5OH, Phenol, Phenylic Acid, Phenic Acid, Phenylic Alcohol.

Obtained by the purification of crude carbolic acid and should be kept in dark amber colored, well-stoppered bottles.

It is a solid at ordinary temperature, and is often colored pinkish or brown upon exposure to light and air, which W. von Hanko (Pharm. Post. xxviii p. 325) claims is due to the presence of certain ammonium compounds, atmospheric dust, the metal of containers and by direct sunlight. A slight discoloration does not interfere with any of its medicinal uses.

It deliquesces on exposure to the air and becomes liquid. It is inflammable and burns with a reddish flame. Soluble in alcohol and water, and the presence of the smallest proportion of water causes it to liquify. It is customary to add 10% of water or glycerin to the acid for dispensing, being in a more convenient form when liquid.

It is distinguished from creasote by coagulating albumen while creasote does not. Official in the ointment and oleate.

Antidotes : Sulphate of sodium which forms the harmless sulphocarbolate, and evacuation of the stomach.

It is irritant, anæsthetic and disinfectant. Dose: 1 to 3 grains (0.065 to 0.19 gm.).

UNGUENTUM ACIDI CARBOLICI (Ointment of Carbolic Acid).

Contains 5% of carbolic acid mixed with simple ointment. The strength was 10% in the U. S. P. 1880.

GLYCERITUM ACIDI CARBOLICI (Glycerite of Carbolic Acid).

A solution of 20% carbolic acid in glycerin. Dose: 5 to 10 minims (0.3 to 0.6 C.c.).

ACIDUM CARBOLICUM CRUDUM (Crude Carbolic Acid).

Consists of various constituents of coal tar, principally phenol and cresol.

Obtained by the fractional distillation of dead oil of coal tar, and distils over between 165°C (329°F) and 190°C (374°F).

It is used principally as a disinfectant.

ACIDUM CITRICUM (Citric Acid) $H_3C_6H_5O_7, H_2O$.

An organic acid prepared from lemon juice by boiling, neutralizing with calcium carbonate, forming calcium citrate, adding sulphuric acid, forming calcium sulphate which is precipitated, concentrating the solution and crystallizing. Efflorescent in warm air and deliquescent in moist air. Soluble in water and alcohol, but nearly insoluble in chloroform, benzol and benzin.

Incompatible with alkaline solutions, whether pure or carbonated, (converting them into citrates), with earthly and metallic carbonates, most acetates, alkaline sulphides and soap.

Official in the syrup, solution of magnesium citrate, and solution of potassium citrate.

Used in cases of scurvy and as a substitute for lemons in lemonade. Dose: 5 to 30 grains (0.32 to 1.94 gm.).

SYRUPUS ACIDI CITRICI (Syrup of Citric Acid).

Contains citric acid, water, spirit of lemon and syrup.

Used as a refrigerant addition to drinks, especially carbonic acid water. If kept long it is apt to acquire a musty taste and to deposit grape sugar, from the action of the acid on cane sugar.

ACIDUM GALLICUM (Gallic Acid) $HC_7H_5O_5, H_2O$. Trioxybenzoic Acid.

An organic acid obtained from tannic acid, and also by boiling powdered galls with dilute sulphuric acid, straining, cooling, and purifying the crystals obtained by using animal charcoal and repeated crystallization.

Slightly soluble in water, but soluble in alcohol. When heated it is changed into pyrogallic acid and carbon dioxide. It is distinguished from tannic acid in not being precipitated by solutions of gelatin, albumen, etc.

It forms a bluish black precipitate with ferric salts, but does not color or form precipitates with ferrous salts when pure.

It is astringent.

Dose: 5 to 15 grains (0.32 to 0.97 gm.).

ACIDUM LACTICUM (Lactic Acid) $HC_3H_5O_3$, Oxypropionic Acid. Ethidene-lactic Acid.

An organic acid obtained by the lactic fermentation of grape sugar or milk sugar, by means of micro-organisms (lactic bacteria) present in stale cheese and containing 75% by weight of absolute acid and 25% by weight of water.

Unites in all proportions with alcohol and water. Official in syrup of calcium lactophosphate. Used principally in dyspepsia. Dose: $\frac{1}{2}$ to 4 fluid drachms (1.85 to 14.78 C.c.).

ACIDUM OLEICUM (Oleic Acid) $HC_{18}H_{33}O_2$.

An organic acid obtained by passing superheated steam through fats and separating from the solid fats by pressure, or by the saponification of olein.

It becomes darker and absorbs oxygen when exposed to the air. It is a yellowish or brownish, oily liquid, having a peculiar lard-like odor and taste. Insoluble in water, but soluble in alcohol, chloroform, benzol, oil of turpentine, volatile and fixed oils. Used in preparing the oleates.

ACIDUM SALICYLICUM (Salicylic Acid) $HC_7H_5O_3$, Ortho-oxybenzoic Acid.

An organic acid obtained synthetically from carbolic acid, and existing naturally in the oils of wintergreen and birch.

The salts of this acid are incompatible with acids, salts of the metals and many alkaloids (in aqueous solution). Permanent in the air. Slightly soluble in water, but soluble in alcohol.

Its solubility is increased without affecting its antiseptic value in the presence of potassium nitrate, ammonium citrate, sodium sulphite, sodium phosphate; with ferric salts it produces a beautiful violet color.

Used in making artificial oil of wintergreen and salol.

It is antirheumatic, antipyretic, and antiseptic.

METHYL SALICYLAS (Methyl Salicylate) $CH_3C_7H_5O_3$, Artificial Oil of Wintergreen.

Prepared by distilling together a mixture of methyl alcohol, sulphuric acid and salicylic acid, and should be kept in well-stoppered bottles, away from the light.

It is practically identical with the oils of wintergreen and birch. Soluble in all proportions in alcohol, ether, glaciale acetic acid and carbon disulphide.

Incompatible with alkalies. Used as an antirheumatic.

SALOL (Salol) $C_6H_5C_7H_5O_3$, Phenyl Salicylate.

Prepared by heating together carbolic acid and salicylic acid in the presence of a dehydrating agent as phosphorus oxychloride whereby the elements of water are withdrawn and the phenyl group unites with the salicylic acid radical.

Permanent in the air. Insoluble in water, but soluble in alcohol, ether, chloroform, fixed and volatile oils. When taken internally it is split up by the action of the pancreatic juice into 36% of carbolic acid and 64% of salicylic acid.

Used as an internal antiseptic. Dose: 15 to 30 grains (0.97 to 2. gm.).

ACIDUM STEARICUM (Stearic Acid) $HC_{18}H_{35}O_2$.

An organic acid found associated with oleic and palmitic acids in tallow and solid fats principally.

It is separated by distillation with superheated steam. Permanent in the air. Insoluble in water, but soluble in alcohol and ether.

It was made official in the U. S. P., 1890, solely for its use in making glycerin suppositories.

ACIDUM TANNICUM (Tannic Acid) $HC_{14}H_9O_9$. Digallic Acid, Gallotannic Acid.

An organic acid obtained from nutgall by extraction with ether and evaporating on glass plates.

It gradually turns darker when exposed to the air and light. Soluble in water, alcohol and glycerin. Distinguished from gallic acid by precipitating solutions of gelatin, albumen, etc.

Incompatible with salts of the metals, alkaloidal solutions, potassium chlorate, solution of starch, albumen, gluten and gelatin. It enters into the ointment, troche, glycerite and styptic collodion.

It is powerfully astringent. When taken internally it is probably converted into gallic acid before absorption.

Dose: 3 to 10 grains (0.2 to 0.67 gm.).

The term TANNIN is applied to a class of vegetable principles the aqueous solutions of which give blue or green colors or precipitates, with ferric salts, and precipitate solutions of albumen and gelatin.

UNGUETUM ACIDI TANNICI (Ointment of Tannic Acid).

Contains 20% of tannic acid and 80% of benzoinated lard.

It is twice the strength of the U. S. P. 1880 ointment.

TROCHISCI ACIDI TANNICI (Troches of Tannic Acid).

Contain tannic acid, tragacanth, sugar and stronger orange flower water. Each troche contains about ½ grain of tannic acid.

GLYCERITUM ACIDI TANNICI (Glycerite of Tannic Acid).

Contains 20% of tannic acid dissolved in glycerin. Used both externally and internally as an astringent.

Dose: 10 to 40 minims (0.6 to 2.5 C.c.).

ACIDUM TARTARICUM (Tartaric Acid) $H_2C_4H_4O_6$.

An organic acid obtained from ARGOLS (an impure potassium bitartrate which gathers on the inside of wine casks during fermentation) by boiling with water, adding calcium carbonate, which forms calcium tartrate and neutral potassium tartrate remains in solution, which is decomposed by the addition of calcium chloride, and the calcium tartrate decomposed by the addition of sulphuric acid with the formation of tartaric acid in solution and the precipitation of calcium sulphate.

It may contain as impurities, sulphuric acid, lime, lead and copper. Permanent in the air. Soluble in water, alcohol and ether, and nearly insoluble in chloroform, benzol or benzin.

Distinguished from all other acids by forming a crystalline precipitate of potassium bitartrate, when added to a neutral solution of potassium. Official in compound effervescing powder.

Used principally in the preparation of effervescing draughts, and often used as a substitute for citric acid in making lemonade. Dose: 5 to 30 grains (0.32 to 1.94 gm.).

Products of Fermentation.

PEPSINUM (Pepsin).

A proteolytic ferment (one having the power of digesting proteid matter) or enzyme (an unorganized ferment) obtained from the glandular layer of fresh stomachs of healthy pigs, and capable of digesting not less than 3000 times its own weight of freshly coagulated and disintegrated egg albumen.

It slowly attracts moisture when exposed to the air. Slightly soluble in water: more so in water acidulated with hydrochloric acid. Insoluble in alcohol, ether or chloroform.

It is used to supply the place of the natural digestive ferment. Official in saccharated pepsin.

Dose: 10 to 15 grains (0.65 to 0.97 gm.).

PEPSINUM SACCHARATUM (Saccharated Pepsin).

Prepared by mixing 10 grams of pepsin with 90 grams of sugar of milk, and should digest not less than 300 times its weight of freshly coagulated and disintegrated egg albumen.

Pepsin acts on albuminoids in the stomach, forming substances called peptone.

PANCREATINUM (Pancreatin).

A mixture of enzymes naturally existing in the pancreas of warm-blooded animals, usually obtained from the fresh pancreas of the hog.

It digests albuminoids in the stomach and converts starch into sugar. Soluble in water and insoluble in alcohol.

Pancreatin acts in alkaline solution and pepsin in an acid solution. Used as a digestant.

ALCOHOL (Alcohol) C_2H_5OH, Spirit of Wine, Rectified Spirit.

A liquid containing 91% by weight and 94% by volume of ethyl alcohol and about 9% by weight of water, obtained by the fermentation of saccharine fluids and subsequent distillation. Also obtained by the distillation of whiskey.

When an infusion of grain is acted on by a ferment, the starch in the grain splits up into dextrine, maltose and dextrose; the dextrin and maltose then are gradually converted into dextrose (grape sugar) which is then converted into alcohol and carbonic acid gas.

Its specific gravity is .820 at 15°C (59°F). It should be free from foreign odor, which, when present, is due to fusel oil. It is inflammable and burns with a blue flame, without odor or smoke. It mixes in all proportions with water, and when they are mixed in equal volume there is a shrinkage of about 3%. It dissolves phosphorus and sulphur in small quantity, iodine, ammonia, potassium hydrate, sodium hydrate, lithium hydrate (but not the carbonates of these metals), organic vegetable alkalies, urea, tannic acid, tartaric acid, camphor, resins, balsams, volatile oils, soap, ammonium chloride, most of the chlorates ready soluble in water, some nitrates, but none of the metallic sulphides. It dissolves the fixed oils sparingly except castor oil, which is soluble in it.

All deliquescent salts are soluble in alcohol (except potassium carbonate). The efflorescent salts and those either insoluble or sparingly soluble in water, are mostly insoluble in it.

The fusel oil in alcohol is removed usually by digestion with charcoal. On allowing alcohol mixed with one-third of its weight of water to evaporate spontaneously from blotting paper there should be no odor of fusel oil or other foreign odor.

ALCOHOL ABSOLUTUM (Absolute Alcohol) Anhydrous alcohol.

Prepared by distilling alcohol with potassium carbonate and calcium chloride, and should contain not more than 1% of water. Its specific gravity is .797 at 15°C (59°F.).

ALCOHOL DEODORATUM (Deodorized Alcohol) Cologne Spirits.

Prepared by filtering alcohol through animal charcoal and rectifying with potassium permanganate. It should contain about 92.5% by weight, and 95.1% by volume of absolute alcohol. Its specific gravity is .816 at 15°C (59°F).

ALCOHOL DILUTUM (Diluted Alcohol) Proof Spirit.

Prepared by mixing together equal volumes of alcohol and water, and contains 41% by weight or 48.6% by volume of absolute alcohol and about 59% of water.

Its specific gravity is .936 at 15°C (59°F). In the U. S. P. 1880, the dilute alcohol was prepared by mixing equal parts by weight of alcohol and water.

Alcoholic liquors having the specific gravity of .920 are termed in commerce, "Proof Spirit." If lighter than this they are said to be above proof; if heavier, below proof; and the percentage of water or spirit of .825 necessary to be added to any sample of spirit to bring it to the standard of proof, indicates the number of degrees the given sample is above or below proof. Thus, if 100 volumes of a spirit require 10 volumes of water to reduce it to proof it is said to be "10 over proof." On the other hand, if 100 volumes of a spirit require 10 volumes of spirit .825 to raise it to proof, it is said to be "10 under proof."

The degree proof is twice the percentage strength by volume.

RUM is obtained by the fermentation of molasses.

HOLLAND GIN by the fermentation of malted barley and rye meal with hops, and rectifying from juniper berries.

COMMON GIN by the fermentation of malted barley, rye or potatoes, and rectified from turpentine.

Official Rules for making an alcohol of any required lower percentage, from an alcohol of any given higher percentage :—

"1" By Volume:—Designate the volume percentage of the stronger alcohol by "V" and that of the weaker by "v."

Rule:—Mix "v" volumes of the stronger alcohol with pure water to make "V" volumes of product. Allow the mixture to stand until full contraction has taken place, and until it has cooled, then make up the deficiency "V" volumes by adding more water.

Ex. An alcohol of 30% by volume is to be made from an alcohol of 94% by volume :—Take 30 volumes of the 94% alcohol and add enough pure water to produce 94 volumes.

"2" By Weight:—Designate the weight percentage of the stronger alcohol by "W" and that of the weaker by "w."

Rule:—Mix "w" parts by weight of the stronger alcohol with pure water to make 'W" parts by weight of product.

Ex. An alcohol of 50% by weight is to be made from an alcohol of 91% by weight:—Take 50 parts by weight of the 91% alcohol, and add enough pure water to produce 91 parts by weight.

SPIRITUS FRUMENTI (Whiskey).

An alcoholic liquid obtained by the distillation of the mash of fermented grain (usually a mixture of corn, wheat and rye) and at least two years old. It should contain 44% to 50% by weight, or 50% to 58% by volume of alcohol.

When freshly distilled it is colorless, but when kept in casks gradually becomes a brownish color which deepens with time.

It is an excellent alcoholic stimulant, and is often used as an antiseptic to ulcers and wounds.

SPIRITUS VINI GALLICI (Brandy) Spirit of French Wine.

An alcoholic liquid obtained by distilling the fermented unmodified juice of fresh grapes, and at least four years old. It contains from 39%

to 47% by weight or 46% to 55% by volume of alcohol. Often artificially prepared by mixing cognac oil with alcohol and coloring.

It is cordial, stimulant and stomachic.

VINUM ALBUM (White Wine).

An alcoholic liquid, prepared by fermenting the juice of fresh grapes, freed from their stems, seeds and skins, and refers to any of the light colored wines, as sherry, madeira, etc., provided they fill the requirements of the U. S. P. It contains about 10% to 14% by weight, or 12.4% to 17.3% by volume of alcohol.

Used in cases of debility.

VINUM RUBRUM (Red Wine).

An alcoholic liquid prepared by fermenting the juice of fresh grapes in the presence of their stems, seeds and skins. The principal red wines are port and claret, etc. It contains from 10% to 14% by weight, or 12.4% to 17.3% by volume of alcohol.

Used in cases of debility.

GLYCERINUM (Glycerin) $C_3H_5(OH)_3$.

A liquid containing not less than 95% of absolute glycerin obtained by the decomposition of fats and fixed oils in the manufacture of soap, the alkalies uniting with the oily acids and freeing the glycerin; also a by-product in the manufacture of lead plaster.

When exposed to the air it slowly absorbs moisture. Its specific gravity should not be less than 1.250 at 15°C (59°F). Soluble in all proportions in alcohol and water. It possesses extensive properties as a solvent, dissolving bromine and iodine, sulphur iodide, potassium and sodium chlorides, the fixed alkalies, some of the alkaline earths, lime for example, and it increases the solvent powers of water. It is sometimes bleached with chlorine and is liable therefore to contain free chlorine and calcium chloride. It is also liable to contain oxalic and formic acids which render glycerin irritating to the skin; the oxalic acid results from the action of sulphuric acid used in purifying it, and the formic acid from the reaction between oxalic acid and glycerin.

Undiluted glycerin should not be mixed with substances which readily give up oxygen, as potassium permanganate, chromic acid, nitric acid, etc.

It is laxative and preservative. Official in the suppository.

It is the base of the glycerites and is also employed in tinctures, fluid extracts, etc., to retard or prevent precipitation.

SUPPOSITORIA GLYCERINI (Glycerin Suppositories).

Prepared by dissolving sodium carbonate in glycerin, adding stearic acid, heating to expel carbonic acid gas, then pouring into moulds and cooling.

The stearic acid and the sodium carbonate unite to form a soap which keeps the glycerin in suspension. Used as a laxative.

LECTURE NO. 13.

Products of Carbonization.

CARBO ANIMALIS (Animal Charcoal) Bone Black, Ivory Black.

Charcoal obtained by heating to redness bones without the access of air.

Insoluble in water or alcohol. It consists of about 90% of calcium phosphate and 10% of charcoal.

It is used in pharmacy for decolorizing vegetable principles such as gallic acid, quinine, morphine, veratine, etc., and in the arts, principally for clarifying syrup in sugar refining, etc.

CARBO ANIMALIS PURIFICATUS (Purified Animal Charcoal).

Animal Charcoal from which the earthy salts have been removed almost wholly, by treatment with hydrochloric acid. The presence of calcium phosphate and carbonate would do no harm in some decolorizing operations, but in delicate chemical processes, these salts might be dissolved or decomposed. Insoluble in water or alcohol.

CARBO LIGNI (Charcoal) Wood Charcoal, Vegetable Charcoal.

Charcoal obtained from soft wood by exposure to a red heat, without the access of air.

Insoluble in water and alcohol. As ordinarily prepared it contains the incombustible part of the wood, amounting to 1% or 2% which is left as ashes when the charcoal is burned. These may be removed by digesting the charcoal in diluted hydrochloric acid, and afterwards washing it thoroughly with boiling water. It is disinfectant and absorbent.

CARBONEI DISULPHIDUM (Carbon Disulphide) Carbon Bisulphide. CS_2.

A body prepared by the direct combination of carbon (in the form of charcoal) and sulphur, at a moderate red heat.

Partially soluble in water, but soluble in alcohol, ether, chloroform, fixed and volatile oils. It vaporizes rapidly at ordinary temperatures, and is highly inflammable. It is a powerful poison, and is not used internally. Its principal uses are in the extraction of oils from oil-seeds, for the extraction of sulphur from sulphur ores, cleansing wool and recovering the fat, in the manufacture of India rubber goods as a solvent of the caoutchouc, and for the extraction of perfumes.

It has been used externally as a counter irritant and local anæsthetic. It is also a valuable antiseptic.

Ethers.

ÆTHER (Ether) $(C_2H_5)_2O$, Sulphuric Ether, Hydric Ether, Hydrate of Ethylen, Oxide of Ethyl.

A liquid containing 96% by weight of absolute ethyl oxide, prepared by distilling together a mixture of sulphuric acid and alcohol, using the whole of the acid and only a third of the alcohol, the remainder of the alcohol being added as fast as ether distills over, keeping a uniform temperature below 141.1°C (286°F). If the whole of the alcohol were used it would distill over in part unchanged, with the ether. The impure ether is purified by shaking with calcium chloride, which takes out sulphurous acid, then with calcium oxide which removes the water, filtered and redistilled.

Its specific gravity is 0.725 to 0.728 at 15°C (59°F). At a temperature of 160°C (320°F) there would be generated sulphurous acid, heavy oil of wine, olefiant gas, and a large quantity of resino-carbonaceous matter, blackening and rendering thick the residuary liquid, all of them being products arising from the decomposition of a portion of sulphuric acid, alcohol and ether.

Soluble in ten times its volume of water at 15°C (59°F). It mixes with chloroform, alcohol, benzin, benzol, fixed and volatile oils. It is highly volatile and inflammable, and should be kept away from lights and fire, the products of its combustion being water and carbonic acid gas.

The vapor of ether is heavier than air. The strength of the U. S. P. 1880 Ether was 74%.

Ether is an excellent solvent for resins, fats, waxes, alkaloids, oils, etc. It is anæsthetic and narcotic.

When too long kept it undergoes decomposition and is decomposed in part into acetic acid.

The amount of ether required to produce insensibility varies with different persons. While giving ether the finger should be kept on the pulse, and if it becomes feeble or very slow or very rapid, the ether should be removed until the circulation becomes restored. The danger in etherization is rarely through failure of the circulation, but by arrest of the respiration, and the state of the latter function should be closely watched : should it become slow, or shallow, or irregular, the ether should be withdrawn, and if necessary, appropriate measures of relief adopted.

One of the disadvantages in the use of ether, is that vomiting is apt to occur and be severe during the recovery from narcosis. No food should be allowed for some hours before etherization, and a moderate dose of brandy or whiskey should be given at the beginning of the latter process. Diseases of the kidneys strongly contra-indicate the use of ether, chloroform being safer under such circumstances.

SPIRITUS ÆTHERIS (Spirit of Ether).

Prepared by mixing together 325 C.c. of ether with 675 C.c. of alcohol.

Dose : 1 to 3 fluid drachms (3.75 to 11.25 C.c.).

SPIRITUS ÆTHERIS COMPOSITUS (Compound Spirit of Ether). Hoffman's Anodyne.

Contains 325 C.c. of ether, 650 C.c. of alcohol, and 25 C.c. of ethereal oil.

It becomes milky when mixed with water, owing to the separation of the ethereal oil.

It is stimulant, antispasmodic and anodyne.

Dose : 30 minims to 2 fluid drachms (1.85 to 7.5 C.c.).

ESTERS.

ÆTHER ACETICUS (Acetic Ether) Ethyl Acetate. $C_2H_3C_2H_3O_2$.

A liquid composed of 98.5% by weight of ethyl acetate, and 1.5% of alcohol, containing a little water, prepared by distilling together a mixture of sulphuric acid, alcohol and sodium acetate.

Soluble in water and mixes in all proportions with ether, alcohol, fixed and volatile oils. Its specific gravity is .893 to .895 at 15°C (59°F). It is inflammable and burns with a yellowish flame and an acetous odor. It is stimulant and antispasmodic. Dose : 15 to 30 drops (0.9 to 1.9 C.c.).

SPIRITUS ÆTHERIS NITROSI (Spirit of Nitrous Ether) $C_2H_5NO_2$. Sweet Spirit of Nitre.

An alcoholic solution of ethyl nitrite, yielding when freshly prepared and tested by a nitrometer, not less than 11 times its own volume of nitrogen dioxide (NO).

Prepared by distilling together a mixture of sodium nitrite, suphuric acid and alcohol, washing the distillate with ice cold water to remove any alcohol, removing traces of acid with ice cold water in which sodium carbonate has been dissolved, agitating with potassium carbonate to remove the water, and filtering into deodorized alcohol. Ascertain the weight of the nitrous ether filtered into the deodorized alcohol by noting the increase in weight of the tared bottle and contents, and then add enough deodorized alcohol to make the mixture weigh twenty-two times the weight of the nitrous ether added.

Its specific gravity is .820 at 15°C (59°F). When kept for a long time after having been freely exposed to light and air it acquires an acid reaction, but it should not effervesce when a crystal of potassium bicarbonate is dropped into it. It mixes with alcohol and water in all proportions. It is very inflammable and burns with a white flame. Alcohol and water are very often fraudulently added to it.

It is diaphoretic, diuretic and antispasmodic. Dose : 30 to 60 minims (1.9 to 3.75 C.c.).

OLEUM ÆTHEREUM (Ethereal Oil) Light Oil of Wine.

A volatile liquid, consisting of equal volumes of heavy oil of wine and ether. Prepared by mixing alcohol and sulphuric acid together, allowing to stand to allow the lead sulphate which is usually present in commercial sulphuric acid to deposit, because its presence in the retort will cause frothing and distilling. The product of distillation is generally in two layers, one consisting of water holding sulphurous acid in solution, and the other, of ether containing the heavy oil of wine. After separation, the liquid is exposed to the air for twenty-four hours in order to evaporate the ether off, the oil is then washed with water to remove the sulphuric acid.

The undiluted ethereal is heavy oil of wine, which is liable to change on keeping, and the presence of ether prevents this change. Insoluble in water, but soluble in alcohol.

Official in compound spirit of ether.

On long keeping it does not change except acquiring a brown hue, which does not interfere with its medicinal action.

AMYL NITRIS (Amyl Nitrite) $C_5H_{11}NO_2$, Amylo-nitrous Ether.

A liquid containing 80% of amyl nitrite, prepared by passing a stream of nitrous acid gas through amyl alcohol.

Its specific gravity is from .870 to .880 at 15°C (59°F.).

Almost insoluble in water. Mixes in all proportions with alcohol and ether. It is very volatile and inflammable.

It is a good antidote to hydrocyanic acid poisoning. Used mostly in pearls which are broken and inhaled on a handkerchief. It is a cardiac stimulant. Dose: 3 to 5 drops (0.18 to 0.3 C.c.) by inhalation or the mouth.

SPIRITUS GLONOINI (Spirit of Glonoin) Spirit of Nitroglycerin, Solution of Trinitin.

Prepared by dissolving nitroglycerin in alcohol and contains 1%.

It should be stored in a cool place, away from light and fire, and should be kept and transported in well-stoppered tin cans. Great care should be used in handling, packing or stirring the spirit, since a dangerous explosion may result if any considerable quantity of it be spilled and the alcohol be partially or wholly lost by evaporation.

Caution should be used in tasting it, as even in small quantities it produces a violent headache. It has the same action as the nitrites, only more powerful. It is often used in cases of poisoning by illuminating gas. Dose: 1 to 2 minims (.06 to .12 C.c.).

Nitroglycerin $(C_3H_5(NO_3)_3$ Trinitrin, Glonoin, Glyceryl Trinitate, is made by the action of sulphuric and nitric acids on glycerin.

It is a yellowish liquid of a specific gravity of 1.525 to 1.6, inodorous, of a sweet, pungent and aromatic taste, slightly soluble in water, readily soluble in alcohol, ether and methylated spirit. It does not freeze above—15°C (4°F) when quite pure, but as it occurs in commerce, solidifies by constant exposure to a temperature of 8°C (46.4°F) and assumes the form of long needles, which are dangerous to handle, as they explode violently when broken. The action of nitroglycerin differs from that of amyl nitrite in that its effects are not so prompt and are perceptible for a much longer time.

METHYL SALICYLAS. See page 91.

Aldehydes.

ALDEHYDE (Acetic Aldehyde, Acetaldehyde) Not Official. C_2H_4O.

The word means alcohol deprived of hydrogen (a*l*cohol *dehy*drogenatum) and is prepared by the oxidation of alcohol.

It is a colorless, mobile, inflammable fluid, having a decidedly pungent, ethereal and suffocating odor. It mixes with water, ether and alcohol, and by exposure to the air is converted into acetic acid by the absorption of oxygen. It is a preservative. Locally it is very irritating.

PARALDEHYDUM (Paraldehyde) $C_6H_{12}O_3$.

A polymeric form of ethyl aldehyde (C_2H_4O), prepared by acting on aldehyde with small quantities of sulphuric or hydrochloric acids.

It is a colorless transparent liquid soluble in water, and mixes in all proportions with alcohol, ether, fixed and volatile oils.

It is hypnotic. Dose: ½ to 1½ fluid drachms (1.8 to 5.5 C.c.).

CHLORAL (Chloral) C_2HCl_3O, H_2O, Chloral Hydrate, Hydrate of Chloral.

Prepared by passing chlorine gas into cold alcohol, then heating the alcohol to not over 60°C (140°F) and passing chlorine gas into it until saturated; it is then saturated with sulphuric acid, during which time the hydrochloric acid gas escapes, and then rectifying over calcium carbonate, which removes the acid; it is then made hydrous with water and crystallized.

It is slowly volatilized when exposed to the air. Soluble in water and alcohol. The name chloral was derived from the *chlor*ine and a*l*cohol from which it is made. It is liquefied when triturated with an equal quantity of menthol, camphor, thymol, or carbolic acid.

It is hypnotic. Dose: 15 to 30 grains (1 to 2 gms.).

Antidotes: emetics or stomach pump, atropine, digitalis or strychnine.

Petrolatum Derivatives.

PETROLATUM LIQUIDUM (Liquid Petrolatum) Paraffine Oil.

A mixture of hydrocarbons, chiefly of the marsh gas series, obtained by distilling off the lighter and more volatile portions from petrolatum, and purifying the residue when it has the desired consistence.

It is a colorless or more or less yellowish, oily liquid, transparent, without odor or taste. Insoluble in water, sparingly soluble in cold or hot alcohol, soluble in ether, chloroform, carbon disulphide, oil of turpentine, benzin, benzol, fixed and volatile oils.

Used principally in the arts.

PETROLATUM MOLLE (Soft Petrolatum).

A mixture of hydrocarbons, chiefly of the marsh gas series, obtained by distilling off the lighter and more volatile portions from petrolatum, and purifying the residue when it has the desired melting point.

A fat-like mass, of about the consistence of an ointment, varying from white to yellowish or yellow. It is to be dispensed where petrolatum is ordered, when not otherwise specified.

It is similar to vaseline, cosmoline, etc.

Used as a vehicle for ointments and dressings.

PETROLATUM SPISSUM (Hard Petrolatum).

A mixture of hydrocarbons, chiefly of the marsh gas series, obtained by distilling off the lighter and more volatile portions from petrolatum and purifying the residue when it has the desired melting point.

It is a fat-like mass, of about the consistence of a cerate, varying from white to yellowish or yellow.

BENZINUM (Benzin) Petroleum Benzin, Petroleum Ether.

A purified distillate from American Petroleum, consisting of hydrocarbons, chiefly of the marsh gas series.

Insoluble in water. Soluble in alcohol, ether, chloroform, benzol, fixed and volatile oils. It is very inflammable and should be kept away from lights and fires. It is used as a solvent for fats, resins, oils, etc.

It should not be confounded with Benzene, which is obtained in the distillation of coal tar.

CHLORINE DERIVATIVES.

CHLOROFORMUM (Chloroform) $CHCl_3$.

A liquid consisting of 99% to 99.4% of absolute chloroform, and from 1% to .6% of alcohol, prepared by distilling together a mixture of chlorinated lime, slaked lime and alcohol, shaking the impure chloroform with water, which separates the alcohol, chlorine, and other contaminating substances; the pyrogenous oil is decomposed with sulphuric acid, which is charred and partially changed into sulphurous acid; the acid is removed by shaking with calcium chloride and lime, which also removes the water; it is then redistilled and 1% of alcohol added to prevent decomposition.

Pure chloroform decomposes by the action of light and air into phosgene gas and chlorine, some hydrochloric acid being also formed. The addition of alcohol partially prevents this change. It is volatile and not inflammable. Specific gravity should not be below 1.490 at 15°C (59°F). Partially soluble in water and freely soluble in alcohol ether, benzin, benzol, and the fixed and volatile oils. It is liable to decomposition by sunlight or even diffused daylight, and should therefore be kept in bottles covered with dark paper, in a dark place.

It has extensive solvent powers dissolving caoutchouc, gutta percha, mastic, tolu, benzoin, copal, iodine, bromine, the organic alkalies, fixed and volatile oils, camphor, most resins, and fats. It dissolves sulphur and phosphorus sparingly.

It is anæsthetic and when locally applied, irritant, and often used as a counter irritant and narcotic in neuralgia, colic, etc. Dose: 2 to 5 minims (.12 to .30 C.c.).

The patient should always be in a horizontal position when inhaling chloroform and the moment snoring is produced it should be withdrawn. It should not be administered to persons subject to epilepsy, affected with organic disease of the heart, or predisposed to syncope.

The reason that ether is much safer than chloroform, lies in the fact that chloroform is a direct parylizant to the heart, while ether is stimulant to it. As chloroform accidents are very liable to occur it is necessary to carefully watch the patient. In most cases a peculiar pallor of the face is the first sign of danger. The patient should be placed at an angle of 45 degrees, head downward, or even completely inverting the person, cold air fanned upon the face, cold water poured upon the head, sinapisms to the feet, friction and heat to the body and extremities, and ammonia to the nostrils. If respiration ceases, the tongue should be seized with the artery forceps and pulled forward from the glottis, and artificial respiration vigorously performed. When the patient can swallow, strong alcoholic drinks may be given.

When taken in an overdose by the mouth, the stomach should be emptied as soon as possible, and the treatment the same as in case of narcosis from inhalation.

Official in the water, spirit, emulsion and liniment.

AQUA CHLOROFORMI (Chloroform Water).

Prepared by shaking chloroform with water, allowing it to settle, and pouring off the clear liquid, which contains about $\frac{1}{2}\%$ and is probably saturated.

It is antiseptic. Dose: $\frac{1}{2}$ to 2 fluid ounces (15 to 60 C.c.).

SPIRITUS CHLOROFORMI (Spirit of Chloroform) Chloric Ether, Spirit of Chloric Ether.

Prepared by mixing together 60 C.c. of chloroform and 940 C.c. of alcohol, and contains 6% by volume, while the spirit of the U. S. P. 1880 contained 10% by weight. Dose: 10 to 60 minims (.6 to 3.75 C.c.).

EMULSUM CHLOROFORMI (Emulsion of Chloroform).

Contains 4% of chloroform expressed oil of almond to render the emulsion permanent, tragacanth and water.

The emulsion of the 1880 U. S. P. contained 5%. Dose: 1 to 2 tablespoonfuls (15 to 30 C.c.).

LINIMENTUM CHLOROFORMI (Liniment of Chloroform).

Contains 300 C.c. of chloroform and 700 C.c. of soap liniment.

It is an excellent application in painful affections. As the chloroform rapidly evaporates, it is desirable, in order to obtain its full anodyne effect, to guard against this by using waxed paper, or some other impervious covering.

IODINE DERIVATIVES.

IODOFORMUM (Iodoform) CHI_3.

Prepared by the action of iodine on alcohol in the presence of a carbonated alkali.

It is slightly volatile, even at ordinary temperatures. Slightly soluble in water, but soluble in alcohol, ether, chloroform, benzin, fixed and volatile oils. The odor of iodoform may be masked to a certain extent by the addition of about 4% of menthol.

Incompatible with potassa in solution, being decomposed, yielding potassium formate and iodide.

It is anæsthetic and antiseptic. Official in the ointment.

When iodoform is brought in contact with the living surface of a wound it is slowly decomposed liberating free iodine, which has a distinct influence upon the development of septic organisms.

UNGUENTUM IODOFORMI (Ointment of Iodoform).

Contains 10 grams of iodoform and 90 grams of benzoinated lard.

It is important to have the iodoform in as fine a powder as possible. Owing to the powerful odor of iodoform it is usual to add oil of bitter almonds, or one of the other volatile oils, Balsam of Peru, or a similar substance, to render the odor more bearable.

Benzene Derivatives.

ACETANILIDUM (Acetanilid) $C_6H_5NHC_2H_3O$, Phenyl-acetamide Antifebrin.

An acetyl derivative of anilin prepared by acting on anilin, with glaciale acetic acid.

Permanent in the air. Soluble in water, and very soluble in alcohol.

Acetanilid sometimes exists as impurity in phenacetine and may be detected as follows: By mixing 5 grains of phenacetine with 2 fluid drachms of solution of potash, boiling, adding 5 drops of chloroform, again boiling, if acetanilid be present a penetrating and repulsive odor will appear.

It is antipyretic and is sometimes used as a substitute for iodoform in hard and soft venereal sores. Dose: 10 grains (.65 gm.).

NAPTHALINUM (Napthalin) $C_{10}H_8$.

A hydrocarbon obtained from coal tar by distillation Insoluble in water, but soluble in alcohol, chloroform, carbon disulphide, naptha, and the oils.

It is antiseptic and disinfectant. Dose: 2 to 8 grains (.13 to .52 gm.) best administered in capsules.

Under the name TAR CAMPHOR, it is largely used in place of camphor to prevent the deposition by moths of eggs, in woolen clothing, and of preventing the destruction by insects in Natural History Museums.

NAPTHOL (Napthol) $C_{10}H_7OH$, Beta-napthol.

A phenol occurring in coal tar or obtained artificially from napthalin by the action of sulphuric acid and fusing with alkalies.

Permanent in the air. Sparingly soluble in water, but soluble in alcohol, ether, chloroform, or solutions of the caustic alkalies.

It is antiseptic. Dose: 3 to 4 grains (.2 to .25 gm.).

LECTURE NO. 14.

PHENOLS.

ACIDUM CARBOLICUM. See page 89.

THYMOL (Thymol) $C_{10}H_{14}O$.

A phenol occurring in the volatile oils of Thymus vulgaris, Monarda punctata, and Carum Ajowan.

Slightly soluble in water, but very soluble in alcohol, ether, fixed oils. When triturated with an equal quantity of camphor, menthol or chloral, it liquefies. It is antiseptic. Dose: 2½ grains (0.162 gm.).

CREOSOTUM (Creosote).

A mixture of phenols, chiefly guaiacol and creosol, obtained by the distillation of wood, preferably that of the beech.

When pure it is colorless, but on exposure to the air and light it turns darker and sometimes as dark as brown, and should be kept in amber-colored bottles. Partially soluble in water, but soluble in alcohol, ether, chloroform, benzin, carbon disulphide, acetic acid, and fixed oils.

It is apt to contain eupion and is sometimes adulterated with rectified oil of tar, fixed and volatile oils, and carbolic acid. Distinguished from carbolic acid by not coagulating albumen.

It is antiseptic. Dose: 1 to 3 minims (.06 to .18 C.c.) given in pill, mixture or solution.

Antidotes: Evacuation of the stomach, then ammonia and other stimulants. Official in the water.

AQUA CREOSOTI (Creosote Water). Contains 1% of creosote or 4.8 minims to each fluid ounce.

Dose: 1 to 4 fluid drachms (3.69 to 15 C.c.).

NAPTHOL. See page 104.

PYROGALLOL (Pyrogallol) $C_6H_3(OH)_3$, Pyrogallic Acid.

A triatomic phenol obtained by the dry distillation of gallic acid.

It acquires a grey or darker tint on exposure to light and air, and should be kept in dark amber-colored bottles. Soluble in water, alcohol and ether. Used in skin diseases; but is very poisonous and should therefore be used with great care.

RESORCINUM (Resorcin) $C_6H_4(OH)_2$, Metadioxybenzol.

A diatomic phenol prepared by fusing galbanum, asafoetida, and other resins with potassa; or by the destructive distillation of brazilin; or from the wash or mother liquors obtained in its manufacture from Brazil Wood.

It is isomeric with pyrocatechin and hydroquinone. It acquires a reddish or brownish tint by exposure to light and air, and should be kept in dark amber-colored bottles. Soluble in alcohol, water, ether, glycerin. Slightly soluble in chloroform.

It is antiseptic and antipyretic. Dose: 2 to 5 grains (.13 to .32 gm.).

WAXES.

CERA FLAVA (Yellow Wax) Beeswax.

A peculiar concrete substance, prepared by the hive bee (Apis mellifica), obtained by fusing the honey-comb in hot water, collecting and straining.

Insoluble in water or alcohol, but soluble in boiling alcohol, ether, chloroform, fixed and volatile oils. Often adulterated with paraffin, resin, meal, earth, and other insoluble substances.

Used in the preparation of ointments, cerates, etc. Official in resin cerate, resin plaster, and ointment.

UNGUENTUM (Ointment) Simple Ointment.

Contains 800 grams of lard and 200 grams of yellow wax.

It is emollient.

CERA ALBA (White Wax).

Obtained by bleaching yellow wax by exposure to moisture, air and light.

Insoluble in water and alcohol. Often adulterated with white lead, starch, meal, etc. Official in simple cerate and the compound cerates.

CERATUM (Cerate) Simple Cerate.

Contains 300 grams of white wax and 700 grams of lard.

CETACEUM (Spermaceti).

A peculiar, concrete, fatty substance, obtained from the head of the sperm whale (Physeter marcrocephalus).

Insoluble in water and alcohol, but soluble in ether, chloroform, carbon disulphide, fixed and volatile oils.

Used in many ointments and cerates. Official in simple cerate and rose water ointment.

STARCHES.

AMYLUM (Starch) $C_6H_{10}O_5$, Wheat Starch, Corn Starch.

The fecula of the seed of Zea Mays (Natural Order, Gramineæ).

It is the proximate principle in most plants, especially in various grains, as wheat, rye, barley, oats, rice, maize, etc., in other seeds as peas, beans, chestnuts, acorns, etc., and in numerous rhizomes and tuberous roots, as those of the potato, sweet potato, etc.

Obtained from these substances by reducing to a fine powder, agitating or washing with cold water, straining or pouring off the liquid, and allowing it to stand until the fine powder or fecula which it holds in suspension has subsided, collecting and drying.

Alkalies unite with starch forming soluble compounds which are decomposed by acids, the starch being precipitated. It is precipitated from its solution by lime water and baryta water, forming insoluble compounds with those earths. Insoluble in water, ether or alcohol, but when added to hot water it swells up and forms a soft, semi-transparent paste, or a gelatinous opaque solution, according to the amount of starch used. When acted on with dilute acid or ferments it is changed to DEXTRIN and Dextrose (grape sugar). The most delicate test for starch is tincture of iodine, which gives a blue color.

It is nutrient and demulcent. Official in the glycerite.

GLYCERITUM AMYLI (Glycerite of Starch) Plasma.

Prepared by heating together starch, glycerin and water.

It is liable to absorb moisture and should be kept in well-closed vessels.

Used as a vehicle for other substances to be employed locally.

Cellulose Derivatives.

CELLULOSE or CELLULIN is the woody fibre of plants which form the skeleton or framework for the vegetable tissue. Raw cotton is a pure form of cellulose.

It is a very important and valuable substance obtained from plants. Hemp, Linen and Cotton goods are made from cellulose.

Cellulose is insoluble in water, alcohol, ether, benzin and oils. When treated with strong sulphuric acid or phosphoric acid it is converted into dextrine. If the mixture be diluted with water and heated it produces glucose. It is converted into parchment paper by passing unsized paper (cellulose) through a mixture of sulphuric acid and water and drying. When acted on with nitric acid it is converted into gun cotton.

LIGNIN is the substance which lines the interior of the skeleton or framework of plants.

PYROXYLINUM (Pyroxylin) Soluble Gun Cotton, Colloxylin.

Prepared by macerating cotton in a mixture of nitric and sulphuric acids until when taken out and washed, it is soluble in a mixture of three parts of ether and one part of alcohol by volume.

When cotton is acted on by nitric acid there is produced under different circumstances five nitro-derivatives: Di, tri, tetra, penta, and hexanitro-cellulose.

The hexanitrate is insoluble in alcohol and ether, and is the true explosive gun cotton. The soluble gun cotton is a mixture of the Di, tri, tetra, and penta nitrates, but principally of the tri and tetra nitrates. Insoluble in water and alcohol, but soluble in acetic ether and amyl acetate.

Used in making collodion, and is now extensively used as the basis of transparent varnishes for lacquering wood and metal.

COLLODIUM (Collodion).

Prepared by dissolving gun cotton in a mixture of alcohol and ether, letting stand and decanting the clear liquid. The sediment consists of undecomposed filaments of cotton.

When applied to a dry surface the ether evaporates quickly and a transparent film is left, having remarkable adhesive and contractile properties. It should be kept in well-stoppered bottles, in a cool place, away from lights and fire. Official in flexible and styptic collodion.

COLLODIUM CANTHARIDATUM (Cantharidal Collodion) Blistering Collodion.

Prepared by percolating cantharides with chloroform, distilling off the chloroform, evaporating the residue and mixing with flexible collodion.

It should be kept in a cool place, away from lights and fire. It is epispastic.

COLLODIUM FLEXILE (Flexible Collodion).

Contains Collodion, Canada turpentine, and castor oil. The Canada turpentine and castor oil render it elastic and not so contractile.

COLLODIUM STYPTICUM (Styptic Collodion) Styptic Colloid, Xylostyptic Ether.

Contains tannic acid, ether, alcohol, and collodion.

Sugars.

CASSIA FISTULA (Purging Cassia) Cassia Pulp.

The fruit of Cassia Fistula (nat. ord. Leguminosæ).

Grows in Upper Egypt and India, and contains sugar, gum, a substance analagous to tannin, a little coloring matter, soluble in ether, traces of a principle resembling gluten, and water.

Official in confection of senna.

It is laxative in doses of 1 to 2 drachms (3.9 to 7.8 gm.).

FICUS (Fig).

The fleshy receptacle of Ficus Carica (nat ord. Urticaceæ).

Grows in Italy and France, and contains grape sugar and gum or mucilage. Official in confection of senna.

It is nutrient, laxative and demulcent.

MANNA (Manna).

The concrete, saccharine exudation of Fraxinus Ornus (nat. ord. Oleaceæ).

It grows in Sicily, Calabria and Apulia. During the hot months the juice exudes from the bark and concretes upon its surface, but as the exudation is slow, it is customary to facilitate the process by making deep incisions longitudinally on one side of the trunk.

It melts and takes fire, burning with a blue flame. When pure it is soluble in water and alcohol, and contains a peculiar sweet principle, Mannite (a hexatomic alcohol), a variety of sugar, a yellow nauseous matter, and mucilage.

When kept for a long time, manna acquires a deeper color, softens, and ultimately deliquesces into a liquid, which, upon the addition of yeast undergoes the vinous fermentation.

It is gently laxative. It is said that manna recently gathered is less purgative than that kept for some time. Dose: 1 to 2 ounces (31.1 to 62.2 gm.). Official in compound infusion of senna.

MEL (Honey).

A saccharine secretion deposited in the honey comb by Apis mellifica (class insecta: order, Hymenoptera).

It should contain from 70% to 80% of glucose. When fresh it is liquid, but on keeping, is apt to form a crystalline deposit and to be converted into a soft granular mass. Largely adulterated with artificial glucose, starch, gelatin, etc. It has the same properties as sugar. Used in making clarified honey.

MEL DESPUMATUM (Clarified honey).

Prepared by mixing honey with 2% of paper pulp, heating as long as any scum rises to the top, skimming off and adding water to make up the loss by evaporation and mixing with 5% of its weight of glycerin.

PRUNUM (Prune).

The fruit of Prunus domestica (nat. ord. Rosaceæ).

It grows in the South of France and contains water, nitrogenous material, fat, free acid, sugar, etc. Official in confection of senna.

It is laxative and nutritious.

RUBUS IDÆUS (Raspberry).

The fruit of Rubus Idæus (nat. ord. Rosaceæ). It is cultivated in the United States and contains sugar, acid, etc. Official in the syrup.

SYRUPUS RUBI IDÆI (Syrup of Raspberry).

Prepared by reducing the fruit to pulp, letting stand in a temperature of 20°C (68°F) until a small portion of the filtered juice mixes clean with half its volume of alcohol, separating the juice, filtering, and to every 40 parts by weight of filtrate, add 60 parts by weight of sugar, boil and strain, and keep in well-stoppered bottles, in a cool and dark place.

SACCHARUM (Sugar) $C_{12}H_{22}O_{11}$, Cane Sugar, Sucrose.

The refined sugar from Saccharum Officinarum.

The saccharine principles distinguished by chemists are as follows: Cane sugar, from sugar cane, sugar beet, and sugar maple, having the formula $C_{12}H_{22}O_{11}$: Lactose or milk sugar, from the whey of cows' milk, and Maltose, a product of the action of malt on cereals, both having the same formula $C_{12}H_{22}O_{11}$. Glucose, including Dextrose or grape sugar, Levulose or fruit sugar, and that resulting from the change of starch and starch containing cereals, formula, $C_6H_{12}O_6$. Arabinose, from gum Arabic, having the formula $C_6H_{12}O_6$, Mannite, from manna, and Dulcite, having the formula $C_6H_{14}O_6$.

Permanent in the air. Soluble in water, 175 parts of alcohol, but insoluble in ether, chloroform, and carbon disulphide.

Strong nitric acid with heat acts on sugar-producing oxalic acid. The same acid when weak converts it into saccharic acid. When heated with dilute sulphuric acid it is converted into glucose. With strong sulphuric acid it is charred. Cane sugar unites with alkalies and some of the alkaline earths forming definite compounds which render the sugar less liable to change. It unites with lead monoxide.

It is an aliment and condiment, and often used as a preservative. Official in Syrup, and used in the preparation of syrups, elixirs, mixtures, emulsions, etc.

SYRUPUS (Syrup) Simple Syrup.

Prepared by dissolving sugar in water with the aid of heat, and contains 85%.

When sugar is used containing a blue coloring matter (ultramarine) the syrup is not colorless, and on standing deposits a dark-colored sediment. Specific gravity 1.317 at 15°C (59°F).

Used in the formation of pills, mixtures, etc.

SACCHARUM LACTIS (Sugar of Milk) $C_{12}H_{22}O_{11}$, H_2O, Lactin, Milk Sugar, Lactose.

A peculiar, crystalline sugar obtained from the whey of cows' milk by evaporation, and purified by recrystallization.

It exists in milk to the extent of about 5%. Soluble in water, but insoluble in ether and alcohol, and is not susceptible to vinous fermentation by the direct action of yeast.

Used as a bland article of diet in consumption.

Gums and Mucilaginous Substances.

ACACIA (Acacia) Gum Arabic.

A gummy exudation from Acacia Senegal (nat. ord. Leguminosæ). Grows in Senegal, Turkey, Barbary and India, and contains a soluble portion (arabin) and an insoluble portion (bassorin), also a lime compound, and is sometimes called gummate of calcium. It is bleached by exposure to the air. Soluble in water, but insoluble in alcohol, ether and the oils. When dissolved in water, it forms a thick, glutinous liquid, having a distinctly acid reaction.

Powdered acacia when treated with iodine test solution, is not colored blue (absence of starch) or red (absence of dextrin). It is often adulterated with starch, flour and cheaper gums. If alcohol be added to mucilage of acacia, arabin will be precipitated. When acted on with strong nitric acid it forms nitro-compounds; with dilute nitric acid it forms mucic and saccharic acids, with oxalic and a little tartaric acid; with dilute sulphuric acid and boiling it forms arabinose or arabin sugar. Its aqueous solution is not precipitated by neutral lead acetate, but the basic acetate throws down a gelatinous precipitate, even in very dilute solution.

Its aqueous solution yields a gelatinous precipitate with ferric chloride test solution, or concentrated solution of borax. Solutions of acacia do not ferment on the addition of yeast, saliva, or gastric juice, but fermentation takes place upon the addition of chalk and cheese, forming lactic acid and alcohol, but not mannite or glycerin. Acacia may be distinguished from dextrine as follows:

1—Gum contains no dextro-glucose while dextrine does, which may be detected by Fehling's Solution.

2—Gum contains a lime compound, and its solutions are rendered milky by the addition of oxalic acid, while the acid does not affect a solution of dextrine. It is demulcent. Official in the mucilage and compound chalk powder.

MUCILAGO ACACIÆ (Mucilage of Acacia).

Prepared by washing acacia with water and draining, then dissolving in water and straining to remove foreign substances often mixed with acacia, and contains 34% of acacia.

A mucilage that will keep for months is prepared by using tolu water instead of water. The mucilage is rendered much more adhesive by the addition of one part aluminum sulphate to one hundred and twenty-five parts of mucilage. It is official in the syrup.

SYRUPUS ACACIÆ (Syrup of Acacia).

Prepared by mixing together 25 C.c. mucilage of acacia and 75 C.c. of syrup.

Useful in the preparation of mixtures, pills and troches. It is demulcent.

ALTHÆA (Marshmallow).

The root of Althæa officinalis (nat. ord. Malvaceæ).

Grows in the United States and Europe, and contains mucilaginous matter, starch, saccharine matter.

Pieces which are woody, discolored, mouldy, of a sour or musty smell, or of a sourish taste should be rejected.

It is demulcent. Official in the syrup.

SYRUPUS ALTHÆÆ (Syrup of Marshmallow).

Prepared by washing marshmallow cut into small pieces, with water, macerated in a mixture of alcohol and water, straining, dissolving sugar in the liquid and adding glycerin.

It should be kept in well stoppered bottles, in a cool place, completely filled. The glycerin and alcohol delay fermentation.

Dose: 1 drachm to $\frac{1}{2}$ fluid ounce (3.75 to 15 C.c.).

CETRARIA (Iceland Moss).

The whole plant of Cetraria islandica (class Lichenes).

Grows in Iceland and also on the mountains and sandy plains of New England, and contains Cetrarin, a peculiar bitter principle, chlorophyll, uncrystallizable sugar, gum, extractive, and a peculiar starch like principle.

The Lapps and Icelanders use powdered cetraria to make bread or boil it with milk, after partially freeing it from bitter principle by repeated maceration in water, the gum and starch which it contains being sufficiently nutritive to serve as food.

Official in the decoction. It is demulcent, nutritious and tonic.

DECOCTUM CETRARIÆ (Decoction of Cetraria).

Prepared by washing cetraria, and then boiling with water.

CHONDRUS (Irish Moss) Carragheen.

The whole plant of Chondrus crispus and Gigartina mamillosa (class. Algæ).

Grows in Europe and the United States, and contains carragheenin, water, mineral matter, albuminoids, and cellulose. Carragheenin has been used as a substitute for acacia but may be distinguished from it as follows: When dissolved in water, carragheenin does not give a precipitate with alcohol. Distinguished from starch by not turning blue with iodine test solution. It is nutritive and demulcent.

DEXTRINE (not official), is a substance resembling gum in appearance and properties, prepared by the action of dilute acids or diastase on starch.

Soluble in water, hot or cold, and forms a mucilaginous solution from which it is precipitated by alcohol. It differs from gum arabic by not affording mucic acid by the action of nitric acid. It may be distinguished from gum by the taste and smell of potato oil which it possesses.

SASSAFRAS MEDULLA (Sassafras Pith).

The pith of Sassafras variifolium (nat. ord. Laurineæ).

Grows in New England and the Southern States, and contains mucilaginous matter in large quantities.

Official in mucilage.

MUCILAGO SASSAFRAS MEDULLÆ (Mucilage of Sassafras Pith).

Prepared by macerating sassafras pith in water for three hours and straining, and should be freshly prepared when wanted for use.

Used as an application to the eyes in conjunctivitis, and also ad libitum in febrile diseases.

TRAGACANTHA (Tragacanth).

A gummy exudation from Astragulus gummifer, and other species of Astragulus (nat. ord. Leguminosæ), growing in Asia Minor, and consists principally of bassorin, with a small quantity of arabin.

When treated with water, it swells up and forms a gelatinous mass. Insoluble in alcohol. The gelatinous mass is tinged blue by the addition of iodine test solution, but the fluid portion is precipitated by the addition of alcohol, and is not colored blue by iodine test solution. It is often adulterated with worthless gums.

It is demulcent. Official in the mucilage.

LECTURE NO. 15.

MUCILAGO TRAGACANTHÆ (Mucilage of Tragacanth).

Prepared by boiling together a mixture of glycerin and water, adding tragacanth, macerating twenty-four hours, adding sufficient quantity of water and straining.

When kept it is liable to undergo decomposition, and become offensive, and the addition of a little carbolic acid will prevent this change.

Used chiefly in making pills and troches.

ULMUS (Elm) Slippery Elm.

The inner bark of Ulmus fulva (nat. ord. Urticaceæ).

Grows in the United States, and contains mucilaginous matter in large quantities.

Its mucilage is precipitated by solutions of lead acetate and subacetate, but not by alcohol. Ground elm is usually adulterated with substances containing starch. Pure elm should contain no starch.

It is demulcent. Official in the mucilage.

MUCILAGO ULMI (Mucilage of Elm).

Prepared by digesting elm with water on a water bath for one hour, and straining.

It should be prepared fresh when wanted for use.

FIXED OILS AND FATS.

AMYGDALA DULCIS (Sweet Almond).

The seed of Prunus Amygdalus, variety, dulcis (nat. ord. Rosaceæ).

Grows in Persia, Barbary and Syria, and contains fixed oil, albumen, uncrystallizable sugar, gum, etc.

Used as a nutrient and demulcent. Official in the syrup and emulsion.

EMULSUM AMYGDALÆ (Emulsion of Almond) Almond Mixture, Milk of Almond.

Prepared by blanching sweet almonds, adding sugar and powdered acacia, beating together and adding water.

The acacia is used to assist in the suspension of the insoluble ingredients of the almonds. The emulsion is not permanent.

Used mostly as a vehicle. Dose: 2 to 8 ounces (60 C.c. to 236 C.c.).

SYRUPUS AMYGDALÆ (Syrup of Almond) Syrup of Orgeat.

Prepared by rubbing bitter and sweet almonds, previously blanched in a mortar, with sugar and water to a smooth paste, adding orange flower water, expressing and straining, repeating, dissolving sugar in the strained liquid, and adding syrup. It should be kept in well-stoppered, well-filled bottles, in a dark place.

Used as a demulcent and nutritive, and may be used as an agreeable vehicle for strong remedies.

OLEUM AMYGDALÆ EXPRESSUM (Expressed Oil of Almonds).

A fixed oil obtained from Bitter or Sweet Almond by expression.

Slightly soluble in alcohol; soluble in ether and chloroform. Often adulterated with poppy oil or other drying oils of less value. Used for the same purpose as olive oil. Official in ointment of rose water and phosphorated oil. Dose: 1 fluid drachm to 1 fluid ounce (3.75 to 30 C.c.).

UNGUENTUM AQUÆ ROSÆ (Ointment of Rose Water) Cold Cream.

Contains spermaceti, white wax, expressed oil of almond, stronger rose water, and borax to whiten it.

It is a useful application for chapped hands, and is liable to become rancid when kept long, and the water will separate by exposure.

ADEPS (Lard) Prepared Lard, Prepared Hogs' Lard.

The prepared internal fat of the abdomen of the hog (Sus Scrofa, class, Mammalia).

Purified by washing with water, melting and straining. It is liable to contain common salt, which renders it unfit for pharmaceutical purposes. This may be detected by boiling the lard with water, and adding a little test solution of silver nitrate, which precipitates silver chloride if salt be present. Lard is often adulterated with water, cotton-seed oil, etc.

Pure lard remains perfectly white on standing, while if adulterated with cotton-seed oil it assumes a more or less olive brown color, according to the amount of adulterant used, 1% causing a perceptible change.

Insoluble in water, but slightly soluble in alcohol. Soluble in ether, benzin, carbon disulphide, and chloroform. When melted it readily unites with wax and resin. It consists of olein, palmitin and stearin. Exposed to the air it absorbs oxygen and becomes rancid, which may be prevented by the addition of benzoin.

It is emollient, and is also used in the preparation of ointments and cerates.

ADEPS BENZOINATUS (Benzoinated Lard).

Prepared by melting lard and suspending benzoin in it, enclosed in a muslin bag, heating for two hours and straining. It is preferable to lard in that it will keep without becoming rancid.

ADEPS LANÆ HYDROSUS (Hydrous Wool Fat) Lanoline.

The purified fat of the wool of sheep (Ovis Aries, class, Mammalia).

Sheep's wool contains on an average 45% of its own weight of fat. It is a mixture of ethers of cholesterin, with the several fatty acids contained in ordinary fats. Insoluble in water, but mixes with twice its weight of it. Sparingly soluble in alcohol, but soluble in ether and chloroform.

Used as a base for ointments, for the reason that it is more readily absorbed than lard.

OLEUM GOSSYPII SEMINIS (Cotton-Seed Oil).

A fixed oil expressed from the seed of Gossypium herbaceum and other species of Gossypium (nat. ord. Malvaceæ).

Usually decolorized by filtering through animal charcoal. Sparingly soluble in alcohol, but soluble in ether, chloroform, and carbon disulphide.

Used extensively as an adulterant of other oils, and as a substitute for olive and almond oil.

LINUM (LINSEED) Flaxseed.

The seed of Linum usitatissimum (nat. ord. Lineæ).

Grows in the United States and almost all over the world, and contains fixed oil, wax, resin, extractive, tannin, gum, etc.

The ground seeds are called linseed meal, which is used in making emollient poultices, and is sometimes adulterated with corn meal, or other meals, containing starch, but these may be easily detected by the iodine test.

Linseed is demulcent and emollient.

OLEUM LINI (Linseed Oil).

A fixed oil expressed from linseed without the use of heat.

When exposed to the air it gradually thickens, and acquires a strong odor and taste, and if spread on a glass plate in a thin layer, and allowed to stand in a warm place, it is gradually converted into a hard, transparent, resin-like mass. It is sometimes adulterated with cheap or crude cod liver oil. Soluble in ether, chloroform, carbon disulphide and benzin.

Official in lime liniment and soft soap. Dose: 1 fluid ounce (30 C.c.).

SAPO MOLLIS (Soft Soap) Green Soap.

Prepared by adding potassa previously dissolved in water to linseed oil adding alcohol, and heating. Soluble in hot water and hot alcohol. As made in this country it is a soft, unctuous mass, of a yellowish brown or brownish yellow color. The German green soap is prepared by saponifying linseed, rape seed, or other vegetable oils with various other refuse oils, usually including fish oils, an excess of potash, and soda.

The color of this soap is green, probably due to the chlorophyll in the impure vegetable oils used.

Green soap is used in diseases of the skin, and is official in the liniment.

LINIMENTUM SAPONIS MOLLIS (Liniment of Soft Soap).

Prepared by mixing oil of lavender flowers with alcohol, dissolving soft soap in this, filtering and adding water.

This preparation is the Tincture of Green Soap of the U. S. P. 1880, and is always used externally.

OLEUM MORRHUÆ (Cod Liver Oil).

A fixed oil expressed from the fresh livers of Gadus Morrhua or other species of Gadus (class, Pisces.).

It contains about 70% of olein and 25% of palmitin with some stearin, gaduin and gadic acid.

Often adulterated with other fixed oils. Strong nitric acid when agitated with cod liver oil, immediately causes a pinkish or rose-red color, which soon becomes brown; while no such an effect is produced on other animal or vegetable oils. It is scarcely soluble in alcohol, but soluble in ether, chloroform and carbon disulphide.

Used principally in rheumatic diseases and phthisis. Dose: 1 tablespoonful (14.7 C.c.)

OLEUM OLIVÆ (Olive Oil) Sweet Oil.

A fixed oil expressed from the ripe fruit of the Olea Europæ (nat. ord. Oleaceæ).

Grows in Spain, South of France and Italy, and contains about 72% of olein and about 28% of palmitin. It is largely adulterated with and substituted by cotton-seed oil. Soluble in chloroform, carbon disulphide and ether, but is only partially soluble in alcohol.

It is solidified by nitrous acid and mercuric nitrate, and is converted into a peculiar fatty substance called Elaidin. When exposed to the air, olive oil is apt to become rancid, and acquire a disagreeable smell, a sharp taste and a thicker consistence.

Official in lead plaster and castile soap.

It is nutritious and mildly laxative. Dose: 1 to 2 fluid ounces (30 to 60 C.c.).

SAPO (Soap) Castile Soap, Hard Soap, White Castile Soap.

Prepared by boiling together olive oil and solution of soda. Fats and fixed oils contain one liquid principle, Olein (Glyceryl Oleate) and two solid principles, Palmitin (Glyceryl Palmitate) and Stearin (Glyceryl Stearate): when these are decomposed by salifiable bases, the three corresponding acids, Oleic, Palmitic, and Stearic, are formed, which unite with the bases to form the soap, therefore soaps are mixtures of oleates, palmitates and stearates of the various bases.

When olein, palmitin and stearin are decomposed in the above named manner into the acids, the glyceryl being freed, unites with the elements of water to form glycerin (C_3H_5, glyceryl; $C_3H_5(OH)_3$ glycerin) which is a by-product in the manfacture of soaps. Soaps are divided into two classes, soluble and insoluble. The soluble soaps are combinations of the fatty acids with soda, potassa or ammonia; the insoluble soaps consist of the same fatty acids united with earths and metallic oxides. The term "soap" is applied to the soluble soaps. The more palmitin and stearin they contain the darker they are, hence sodium stearate is the hardest soap and the least soluble, while potassium oleate is the softest and most soluble. Soap is soluble in water and more so in alcohol and hot water. It is sometimes adulterated with lime, silica, gypsum, heavy spar, pipe clay, and sodium sulphate, these substances being detected by not dissolving in alcohol. Soap contains about 21% of water.

It is laxative and antacid. Official in the liniment and plaster. In pharmacy it is sometimes used to give consistence to pills, care being taken not to unite with any substance which may be decomposed by it.

In case of poisoning it is employed as a counter poison for the mineral acids. Dose: 5 to 30 grains (.33 to 1.95 gm.).

LINIMENTUM SAPONIS (Soap Liniment) Camphorated Tincture of Soap, Liquid Opodeldoc.

Contains soap, camphor, oil rosemary, alcohol and water.

The soap should be in the form of a fine powder on account of the water it contains not varying to any great extent.

EMPLASTUM SAPONIS (Soap Plaster).

Contains soap in fine powder, lead plaster and water.

It is discutient.

OLEUM RICINI (Castor Oil).

A fixed oil expressed from the seed of Ricinus communis (nat. ord. Euphorbiaceæ).

Grows in East Indies, North Africa, West Indies, and is cultivated in various parts of the world. It contains Ricinolein, palmitin, stearin and myristin. When exposed to the air it thickens, without becoming opaque. Soluble in all proportions in cold absolute alcohol (different from other fixed oils), also in an equal weight of alcohol, and in ether. The seeds contain besides fixed oil, gum, starch and albumen.

It is a mild and speedy cathartic. Dose: $\frac{1}{2}$ fluid ounce (15 C.c.).

OLEUM SESAMI (Oil of Sesamum) Benne Oil, Teel Oil.

A fixed oil expressed from the seed of Sesamum Indicum (nat. ord. Pedaliaceæ).

It grows in the East Indies and United States, and contains about 76% of olein, with palmitin, stearin, and myristin.

In large doses it is laxative.

OLEUM THEOBROMATIS (Oil of Theobroma) Cacao Butter.

A fixed oil expressed from the seed (or chocolate nut) of Theobroma Cacao (nat. ord. Sterculiaceæ).

It grows in Mexico, West Indies and South America, and contains stearin, palmitin and olein, with the glycerides probably of arachic and lauric acid. Soluble in ether and chloroform.

Used principally in making suppositories, on account of its low fusing point, and power of not becoming rancid.

OLEUM TIGLII (Croton Oil).

A fixed oil expressed from the seed of Croton Tiglium (nat. ord. Euphorbiaceæ).

It grows in Hindoostan, Ceylon, the Moluccas, and other parts of India, and contains the glycerides of stearic, palmitic, myristic, lauric and oleic acid; there are also present in the form of glycerin ethers, the more volatile acids as formic, acetic, isobutyric, and isovalerianic.

When fresh it is soluble in about 60 parts of alcohol and the solubility increases with age. Soluble in ether, chloroform, carbon disulphide, fixed and volatile oils.

It is often adulterated with other volatile oils.

It is a powerful drastic purgative. Dose: 1 to 2 drops (.06 to .12 C.c.).

SEVUM (Suet) Prepared Suet, Mutton Suet.

The internal fat of the abdomen of the sheep, Ovis Aries (class. Mammalia), purified by melting and straning.

Insoluble in alcohol and water. Soluble in ether and benzin. Sometimes used in conjunction with lard as a base for ointments.

VOLATILE OILS.

Volatile oils are divided into four classes namely:

1—Terpenes, which are hydrocarbons, consisting of carbon and hydrogen, and usually have the formula $C_{10}H_{16}$.

2—Oxygenated Oils, which are hydrocarbons containing oxygen.

3—Sulphurated Oils, which contain sulphur.

4—Nitrogenated Oils, which contain hydrocyanic acid: otherwise nitrogen is never contained in a volatile oil.

Volatile oils should be kept in well-stoppered bottles, in a cool place, away from the light.

ABSINTHIUM (Wormwood).

The leaves and tops of Artemisia Absinthium (nat. ord. Compositæ), growing in Eurpoe, and cultivated in New England. It contains volatile oil, (consisting of a terpene and absinthinol), a little nitrogenous matter, a bitter resinous substance, chlorophyll, albumen, starch, saline matters and lignin.

The old Salt of Wormwood (sal absinthii) was impure potassium carbonate, made from the ashes of the plant.

It is a stomachic tonic. Dose: 1 to 2 scruples (1.3 to 2.6 gm.).

ALLIUM (Garlic).

The bulb of Allium sativum (nat. ord. Liliaceæ), growing in Italy, Sicily, and the South of France.

It contains Volatile Oil (consisting of allyl sulphide $(C_3H_5)_2S$), mucilage, albumen, fibrous matter and water. Water, alcohol, and vinegar extract the virtues of garlic.

It is a stimulant and condiment. Dose: ½ to 2 drms. (1.95 to 7.8 gm.). Official in the syrup.

SYRUPUS ALLII (Syrup of Garlic).

Prepared by macerating garlic with diluted acetic acid, expressing, macerating the residue with more dilute acetic acid, again expressing, mixing the liquids, filtering, adding sugar and shaking until dissolved, and making up to the required quantity with dilute acetic acid.

Dose: 1 teaspoonful (3.75 C.c.).

OLEUM AMYGDALÆ AMARÆ (Oil of Bitter Almond).

A volate oil obtained from bitter almond by maceration with water and subsequent distillation, and should be kept in small, well stoppered bottles, away from the light.

It is heavier than water. Soluble in alcohol and ether, and is partially soluble in water. When fresh it is neutral to litmus, but when kept for some time assumes an acid reaction, due to the formation of benzoic acid.

It contains hydrocyanic acid, a small quantity of benzoic acid, and a concrete principle called benzoin, $C_{14}H_{12}O_2$. The oil on standing, deposits benzoic acid, which does not pre-exist in it, but results from the absorption of oxygen.

Official in the water and spirit.

Its action is the same as dilute hydrocyanic acid. Dose : $\frac{1}{4}$ to 1 drop, (0.016 to 0.06 C.c.).

ARTIFICIAL OIL OF BITTER ALMOND (Oil of Mirbane) Nitrobenzol, Nitro-benzene, $C_6H_5NO_2$ is prepared by acting on benzol (benzene) with fuming nitric acid.

It is used considerably in confectionery, soaps, etc. The true oil has a peculiar odor which the artificial oil does not have. The specific gravity of the true oil is 1.069 to 1.070, and boiling point 180°C (356°F), while that of the artificial oil is 1.86, and boiling point 213°C (415°F).

The following test has been proposed by the late Prof. J. M. Maisch for detecting nitro-benzine in oil of bitter almond :

Dissolve one drachm of suspected oil in 2 or 3 drachms of alcohol, add 15 grains of pure fused caustic potassa, heat for a few minutes so as to dissolve the potassa, reduce the liquid to one-third, and then set aside to cool. If the oil be pure, it will remain liquid, while if nitro-benzene be present there will, after cooling, be a crystalline deposit, proportionate to the amount of adulteration.

AQUA AMYGDALÆ AMARÆ (Bitter Almond Water).

Prepared by dissolving 1% of the oil in water by agitation.

Dose : 1 to 2 teaspoonfuls (3.75 to 7.5 C.c.).

SPIRITUS AMYGDALÆ AMARÆ (Spirit of Bitter Almond).

Contains 1% of the oil dissolved in alcohol, with the addition of a small quantity of water.

Used principally as a flavoring.

ANISUM (Anise).

The fruit of Pimpinella Anisum (nat. ord. Umbelliferæ), growing in Egypt, Spain, Romagna, Italy, Russia, and the Levant.

It contains Volatile Oil, upon which its virtues depend, and a fixed oil.

The Spanish fruit is smaller than the German or French, and is usually preferred ; the Russian is very short.

It is aromatic and carminative. Dose : 20 to 30 grains (1.3 to 1.95 gm.).

OLEUM ANISI (Oil of Anise).

A volatile oil distilled from Anise, which should be kept in well-stoppered bottles, protected from light, and if it has separated into a liquid and solid portion, should be completely liquefied by warming before being dispensed.

Soluble in alcohol. It absorbs oxygen from the air, and becomes less disposed to concrete. It should not assume a blue or brownish color on the addition of a drop of ferric chloride test solution (absence of some volatile oils containing phenols).

It contains principally Anethol $C_{10}H_{12}O$ and a terpene $C_{10}H_{16}$. Often adulterated with spermaceti, wax, or camphor; the first two may be detected by their insolubility in alcohol, and the last by its odor.

Official in the water, spirit, compound spirit of orange, and camphorated tincture of opium. Dose: 5 to 15 drops (0.3 to 0.9 C.c.).

AQUA ANISI (Anise Water).

Prepared by triturating oil of anise with precipitated phosphate of calcium, adding water and filtering. Used as a vehicle.

SPIRITUS ANISI (Spirit of Anise).

A solution of the oil in deodorized alcohol, containing 10%.

It is stomachic and carminative. Dose. 1 to 2 fluid drachms (3.75 to 7.5 C.c.).

ANTHEMIS (Anthemis) Roman or English Chamomile.

The flower heads of Anthemis nobilis (nat. ord. Compositæ), growing in France, Germany, Italy, and in some parts of the United States.

It contains Volatile Oil, upon which its virtues depend.

It is a mild tonic. Dose: 30 to 60 grains (1.95 to 3.9 gm.).

AURANTII AMARI CORTEX (Bitter Orange Peel).

The rind of the fruit of Citrus vulgaris (nat. ord. Rutaceæ), growing in Florida and the West Indies.

It contains Volatile Oil, a crystalline acid, a non-crystalline resinous body, hesperidin, iso-hesperidin (a crystalline glucoside isomeric with hesperidin) aurantianarin (a new glucoside to which the bitterness is due).

It is a mild tonic, carminative, and stomachic. Dose: ½ to 1 drachm (1.95 to 3 9 gm.). Official in the fluid extract and tincture.

EXTRACTUM AURANTII AMARI FLUIDUM (Fluid Extract of Bitter Orange Peel).

Prepared in the usual manner from bitter orange peel No. 40, using 600 parts of alcohol, and 300 parts of water as the menstruum.

It is tonic. Dose: 15 to 30 minims (0.9 to 1.9 C.c.).

TINCTURA AURANTII AMARI (Tincture of Bitter Orange Peel).

Prepared by percolating bitter orange peel No. 30, with a mixture of 600 parts of alcohol and 400 parts of water.

Dose: 1 to 2 fluid drachms (3.7 to 7.5 C.c.).

LECTURE NO. 16.

AURANTII DULCIS CORTEX (Sweet Orange Peel).

The rind of the fresh fruit of Citrus Aurantium (nat. ord. Rutaceæ).
Its composition is the same as that of the bitter orange.

It is aromatic. Official in the syrup and tincture.

SYRUPUS AURANTII (Syrup of Orange).

Prepared by boiling sweet orange peel with alcohol, filtering, triturating with sugar and precipitated calcium phosphate, adding water, filtering, dissolving the remainder of the sugar in the filtrate.

Used as a flavoring agent.

TINCTURA AURANTII DULCIS (Tincture of Sweet Orange Peel).

Prepared by macerating and percolating sweet orange peel, cut into small pieces, with alcohol.

OLEUM AURANTII CORTICIS (Oil of Orange Peel).

A volatile oil obtained by expressing the fresh peel from either the bitter or sweet orange, or by putting the scrapings into hot water, depressing the pulp beneath, and skimming off the oil as it rises, or by distillation.

It contains Hesperidine ($C_{10}H_{16}$), and a resinous body. Soluble in alcohol, and carbon disulphide. When kept for some time it should not develop a terebinthinate odor and taste (absence of oil of turpentine or of other oils containing pinene). It is difficult to preserve.

Official in the spirit, compound spirit, and spirit of myrcia.

SPIRITUS AURANTII (Spirit of Orange).

A solution of the oil in deodorized alcohol, containing 5% of the oil.

Used for flavoring purposes.

SPIRITUS AURANTII COMPOSITUS (Compound Spirit of Orange).

Contains oil of orange peel, oil of lemon, oil of coriander, oil of anise, and deodorized alcohol.

It should be kept in completely filled, well-stoppered bottles, in a cool and dark place.

Official in aromatic elixir.

ELIXIR AROMATICUM (Aromatic Elixir).

Prepared by adding compound spirit of orange to deodorized alcohol, adding syrup and distilled water, filtering through precipitated calcium phosphate, washing the filter with a mixture of one volume of deodorized alcohol and three volumes of distilled water to make the required quantity.

Official in elixir of phosphorus.

OLEUM AURANTII FLORUM (Oil of Orange Flowers) Oil of Neroli.

A volatile oil distilled from the fresh flowers of the bitter orange, and should be kept in well-stoppered bottles, in a cool place, away from the light, and contains a fragrant hydrocarbon $C_{10}H_{16}$, and a crystalline solid neroli camphor.

Used principally in cologne water, perfumes, etc.

AQUA AURANTII FLORUM FORTIOR (Stronger Orange Flower Water).

Water saturated with the oil of fresh orange flowers, obtained as a by-product in the distillation of oil of orange, and should be kept in loosely-stoppered bottles, in a dark place.

Used in making orange flower water.

AQUA AURANTII FLORUM (Orange Flower Water).

Prepared by mixing together equal parts of stronger orange flower water and water.

Nitric, hydrochloric or sulphuric acids color the oil in the water a rose tint.

Used on account of its agreeable odor. Official in the syrup.

It becomes red on exposure to light.

SYRUPUS AURANTII FLORUM (Syrup of Orange Flowers).

Prepared by dissolving sugar in orange flower water, without the aid of heat.

Used principally for flavoring.

Dose: 1 fluid drachm (3.75 C.c.).

OLEUM BERGAMOTTÆ (Oil of Bergamot).

A volatile oil obtained by expression from the rind of the fresh fruit of Citrus Bergamia (nat. ord. Rutaceæ).

It should be kept in well-stoppered bottles, in a cool place, away from the light.

It contains several hydrocarbons of the formula $C_{10}H_{16}$, a solid, greasy portion called bergaptene, or bergamot camphor, and linalool acetate $C_{10}H_{17}$, which is decomposed by steam distillation. Soluble in alcohol and glaciale acetic acid.

Distinguished from oils of orange and lemon by being soluble in solution of potassa and forming with it a clear liquid.

Used principally as a perfume.

OLEUM BETULÆ VOLATILE (Volatile Oil of Birch).

A volatile oil obtained by distillation from the bark of Betula lenta (nat. ord. Betulaceæ).

It is identical with methyl salicylate, and nearly identical with oil of wintergreen, and should be kept in well-stoppered bottles, away from the light. It consists entirely of methyl salicylate.

It has the same properties as oil of wintergreen. Dose: 5 to 30 minims (0.3 to 1.8 C.c.).

BUCHU (Buchu).

The leaves of Barosma betulina, and Barosma crenulata (nat. ord. Rutaceæ), growing in Europe.

It contains Volatile Oil (consisting of eleopten, and a crystalline stearopten, diosphenol $C_{14}H_{22}O_3$). The leaves of Barosma crenata are called short buchu and those of Barosma serratifolia the long buchu. Buchu is distinguished from uva ursi by its leaves having a serrated margin, while those of uva ursi are entire.

According to the late Prof. Bedford the short buchu was found to yield an average of 1.21% of volatile oil, while the long buchu gave only .66%, showing it to be inferior in strength. Official in the fluid extract.

Used as a stimulant diuretic.

EXTRACTUM BUCHU FLUIDUM (Fluid Extract of Buchu).

Prepared from buchu No. 60, by the usual process, using alcohol as the menstruum.

On long keeping it acquires the odor of mint, showing that a change takes place in its volatile oil.

Dose: 30 to 60 minims (1.90 to 3.8 C.c.).

OLEUM CAJUPUTI (Oil of Cajuput).

A volatile oil distilled from the leaves of Melaleuca Leucadendron (nat. ord. Myrtaceæ), growing in India, Phillipines, and Australia.

It consists principally of Cajuputene, and is very volatile and inflammable, burning without any residue. Soluble in alcohol.

It is naturally green, but as found in commerce, sometimes contains copper, either accidentally present or added with a view of imitating the color of the oil. Sometimes adulterated with oil of turpentine, or oil of rosemary.

It is highly stimulant. Dose: 5 to 20 drops (0.3 to 1.25 C.c.).

CALAMUS (Calamus) Sweet Flag.

The rhizome of Acorus Calamus (nat. ord. Aroideæ), growing in Europe, Asia and the United States. The root is sometimes attacked by worms, and deteriorates with keeping.

It contains Volatile Oil (containing a terpene $C_{10}H_{16}$), soft resin, extractive, gum, starch, lignin and water.

Official in the fluid extract. It is a feeble aromatic. Dose: 20 to 60 grains (1.3 to 3.9 gm.).

EXTRACTUM CALAMI FLUIDUM (Fluid Extract of Calamus).

Prepared from calamus No. 60 in the usual way, using alcohol as the menstruum.

Dose: 5 to 15 minims (0.3 to 0.9 C.c.).

CARDAMOMUM (Cardamom).

The fruit of Elettaria repens (nat. ord. Scitamineæ), growing in Malabar.

It contains Volatile Oil (consisting of a terpene $C_{10}H_{16}$, with a small quantity of formic and acetic acids), fixed oil, a salt of potassium mixed with coloring matter, starch, mucilage, etc. Sometimes adulterated with orange seeds, and unroasted grains of coffee.

Official in aromatic powder, tincture, compound tincture, compound tincture of gentian, sweet tincture of rhubarb, compound extract of colocynth. It is aromatic.

TINCTURA CARDAMOMI (Tincture of Cardamom).

Prepared by macerating and percolating cardamom No. 30, with dilute alcohol.

Dose: 1 fluid drachm (3.7 C.c.).

TINCTURA CARDAMOMI COMPOSITA (Compound Tincture of Cardamom).

Contains cardamom, cassia cinnamon, caraway, cochineal, glycerin, and dilute alcohol.

Dose: 1 to 2 fluid drachms (3.7 to 7.5 C.c.).

CARUM (Caraway).

The fruit of Carum Carvi (nat. ord. Umbelliferæ), growing in Europe and cultivated in this country.

It contains Volatile Oil, upon which its virtues depend.

Official in compound tincture of cardamom.

It is stomachic and carminative. Dose 20 to 60 grains (1.3 to 3.9 gm.).

OLEUM CARI (Oil of Caraway).

A volatile oil distilled from Caraway and contains Carvene $C_{10}H_{16}$, and Carvol $C_{10}H_{14}O$.

It becomes brownish with age. Soluble in alcohol.

Used principally as a flavoring agent. Official in compound spirit of juniper. Dose: 1 to 10 drops (0.06 to 0.6 C.c.).

CARYOPHYLLUS (Clove).

The unexpanded flowers of Eugenia aromatica (nat. ord. Myrtaceæ), growing in the Molucca Islands, and contains Volatile Oil, a peculiar tannin, gum, resin, vegetable fibre and water.

Official in compound tincture of lavender, aromatic tincture of rhubarb and wine of opium.

It is a stimulant aromatic. Dose: 5 to 10 grains (0.33 to 0.65 gm.).

OLEUM CARYOPHYLLI (Oil of Clove).

A volatile oil distilled from Clove, and contains two distinct oils, one lighter than water and the other heavier. The heavy oil contains Eugenol $C_{10}H_{12}O_2$ and Caryophyllin (a stearopten) and the light oil has the formula $C_{15}H_{24}$.

It becomes yellowish by exposure, and then reddish brown, and when long kept deposits a crystalline stearopten.

Soluble in alcohol, ether and strong acetic acid. Frequently adulterated with fixed oils, and sometimes with the oils of pimenta and copaiba.

Its properties are those of the clove. Dose: 2 to 6 drops (0.12 to 0.36 C.c.).

CHENOPODIUM (Chenopodium) American Wormseed.

The fruit of Chenopodium ambrosioides (nat. ord. Chenopodiaceæ) growing in the United States and contains Volatile Oil upon which its virtues depend.

It is anthelmintic. Dose: 1 to 2 scruples (1.3 to 2.6 gm.).

OLEUM CHENOPODII (Oil of Chenopodium).

A volatile oil distilled from Chenopodium, and contains two distinct oils, one having the formula $C_{10}H_{16}$ and the other the formula $C_{10}H_{16}O$.

It becomes a deeper yellow and brownish with age. Soluble in alcohol.

It is anthelmintic. Dose: 4 to 8 drops (0.24 to 0.5 C.c.).

CINNAMOMUM CASSIA (Cassia Cinnamon) Cassia, Cassia Bark.

The bark of the shoots of one or more undetermined species of Cinnamon grown in China, (nat. ord. Laurineæ).

It contains Volatile Oil, tannin, mucilage, coloring matter, an acid, and lignin. It is a redder and darker color than the Ceylon, thicker, denser and breaks with a short fracture.

Official in compound tincture of cardamom, compound tincture of catechu, compound tincture of lavender, wine of opium, aromatic tincture of rhubarb.

It is aromatic. Dose: 10 to 30 grains (0.65 to 1.3 gm.).

CINNAMOMUM SAIGONICUM (Saigon Cinnamon).

The bark of the undetermined species Cinnamon (nat. ord. Laurineæ) and takes its name from Saigon, the capital of French Cochin China.

It yields a sweeter and less pungent oil than the cassia.

Its properties are the same as those of the other varieties.

CINNAMOMUM ZEYLANICUM (Ceylon Cinnamon), Cinnamon bark.

The inner bark of the shoots of Cinnamomum Zeylanicum (nat ord. Laurineæ).

It is the finest variety, and its virtues depends on a Volatile Oil.

Official in aromatic powder and the tincture. Its properties are the same as those of cassia.

TINCTURA CINNAMOMI (Tincture of Cinnamon).

Prepared by macerating and percolating Ceylon Cinnamon, No. 40, with a mixture of glycerin, alcohol and water.

By long keeping it is liable to gelatinize. Dose: 1 to 3 or 4 fluid drachms (3.7 to 11.25 or 15 C.c.).

PULVIS AROMATICUS (Aromatic Powder), Compound Powder of Cinnamon.

Contains Ceylon cinnamon, ginger, nutmeg and cardamom deprived of their capsules, and crushed.

Official in fluid extract.

It is stimulant and carminative. Dose: 10 to 30 grains (0.65 to 1.95 gm.).

EXTRACTUM AROMATICUM FLUIDUM (Aromatic Fluid Extract), Fluid Extract of Aromatic Powder.

Prepared from aromatic powder, by the usual process, using alcohol as the menstruum.

Dose: 10 to 20 minims (0.6 to 1.25 C.c.).

OLEUM CINNAMOMI (Oil of Cinnamon), Oil of Cassia.

A volatile oil, distilled from Cassia cinnamon, and should be kept in well-stoppered bottles, in a cool place away from the light.

It consists of cinnamic aldehyde C_9H_8O, which by moderate oxidation yields Cinnamic acid, $C_9H_8O_2$, and by more energetic oxidation yields benzoic acid, $C_7H_6O_2$. Oil of cinnamon is frequently adulterated with oil of clove, which can be detected by the smell or taste, and by the following test:

If the oil be shaken with water and passed through a wetted filter, the clear filtrate should give, with a few drops of basic lead acetate solution, a white turbidity without a yellow color (absence of oil of clove).

If some of the oil be dissolved in alcohol, the subsequent addition of a drop of ferric chloride test solution should produce a brown, but not a green or blue color (absence of oil of clove or carbolic acid).

The oil is soluble in alcohol, and becomes darker and thicker by age and exposure to the air.

It is cordial and carminative. Dose: 1 to 3 drops (0.06 to 0.18 C.c.). Official in the water and spirit.

There is a volatile oil of Ceylon cinnamon in the market, which, when fresh, is of a light yellow color, becoming deeper by age, and finally red.

AQUA CINNAMOMI (Cinnamon Water).

Prepared by triturating oil of cinnamon with precipitated calcium phosphate, adding water and filtering. On standing it is apt to precipitate, owing to gradual oxidation and formation of Cinnamic Acid which is insoluble in water.

SPIRITUS CINNAMOMI (Spirit of Cinnamon).

A solution of the oil in alcohol and contains 10%.

Used as an agreeable aromatic cordial. Dose: 10 to 20 drops (0.6 to 1.25 C.c.).

CORIANDRUM (Coriander).

The fruit of Coriandrum sativum (nat. ord. Umbelliferæ) growing in Europe. It contains a Volatile Oil upon which its aromatic smell and taste depend.

Official in confection of senna.

Used as a feeble aromatic. Dose: 20 to 60 grains (1.3 to 3.9 gm.).

OLEUM CORIANDRI (Oil of Coriander).

A volatile oil distilled from coriander, consisting chiefly of Coriandrol, $C_{10}H_{18}O$, which by the removal of a molecule of water, becomes a terpene.

Soluble in alcohol. Often adulterated with rectified oil of orange, which may be detected by its insolubility in 90% alcohol. Official in compound spirit of orange and confection of senna.

It possesses the medical properties of the fruit.

CUBEBA (Cubeb).

The unripe fruit of Piper Cubeba (nat. ord. Piperaceæ) growing in Java and other parts of the East Indies.

It contains Volatile Oil, upon which its properties depend, and Cubebin.

The fruit gradually deteriorates by age, and in powder becomes rapidly weaker, in consequence of its loss of volatile oil, and on this account should be kept whole and powdered when dispensed.

Official in the fluid extract, oleoresin and tincture.

It is stimulant. Dose: 1 to 3 drachms (3.9 to 11.65 gm.).

EXTRACTUM CUBEBÆ FLUIDUM (Fluid Extract of Cubeb).

Prepared from cubeb No. 60, by the usual process, using alcohol as the menstruum.

Dose: 10 to 40 minims (0.6 to 2.5 C.c.).

OLEORESINA CUBEBÆ (Oleoresin of Cubeb).

Prepared by percolating cubeb No. 30 with ether, and evaporating off the ether.

It should be kept in well-stoppered bottles, and consists principally of volatile oil and resin, a small portion of cubebin, and waxy matter of cubeb. It deposits, on standing for some time, a waxy and crystalline matter, which should be rejected, only the liquid portion being used.

Official in the troche. Dose: 5 to 30 minims (0.3 to 1.9 C.c.).

TROCHISCI CUBEBÆ (Troches of Cubeb).

Contain oleoresin of cubeb, oil of sassafras, extract of licorice, acacia and syrup.

TINCTURA CUBEBÆ (Tincture of Cubeb).

Prepared by macerating and percolating cubeb No. 30 with alcohol.

Dose: 1 to 2 fluid drachms (3.7 to 7.5 C.c.).

OLEUM CUBEBÆ (Oil of Cubeb).

A volatile oil distilled from cubeb, containing a small amount of hydrocarbon $C_{10}H_{16}$, two oils of the formula $C_{15}H_{24}$ and a stearopten or cubeb camphor.

When exposed to the air it thickens without losing its odor. Soluble in an equal weight of alcohol. Often used in place of the powder. Dose: 10 to 12 drops (0.6 to 0.72 C.c.).

OLEUM ERIGERONTIS (Oil of Erigeron) Oil of Fleabane.

A volatile oil distilled from the fresh flowering herb of Erigeron Canadense (nat. ord. Compositæ), and probably consists of two distinct oils.

By exposure it becomes darker and thicker. Soluble in alcohol. Used in diarrhœa and dysentery. Dose: 10 to 30 minims (0.6 to 1.8 C.c.).

EUCALYPTUS (Eucalyptus).

The leaves of Eucalyptus globulus (nat. ord. Myrtaceæ), growing in Europe and the United States, and collected from the older parts of the tree.

It contains a Volatile Oil, chlorophyll, resin, tannin and inert substances.

Official in the fluid extract.

It is astringent and narcotic.

EXTRACTUM EUCALYPTI FLUIDUM (Fluid Extract of Eucalyptus).

Prepared from Eucalyptus No. 40, in the usual manner, using a mixture of 3 parts of alcohol and 1 part of water as the menstruum.

Dose: 5 to 10 minims (0.3 to 0.6 C.c.).

OLEUM EUCALYPTI (Oil of Eucalyptus).

A volatile oil distilled from the fresh leaves of eucalyptus, and contains Cymene $C_{10}H_{14}$, eucalyptene $C_{10}H_{16}$, a small quantity of a terpene $C_{10}H_{16}$ and Eucalyptol $C_{10}H_{18}O$, upon which its value depends.

Soluble in alcohol, carbon disulphide and glaciale acetic acid.

Dose: 10 to 15 minims (0.6 to 0.9 C.c.).

EUCALYPTOL (Eucalyptol) $C_{10}H_{18}O$.

A neutral body obtained from fhe volatile oil of eucalyptus, and should be kept in well-stoppered bottles, in a cool place, protected from the light.

Soluble in alcohol, carbon disulphide and glaciale acetic acid. It is a stimulant expectorant. Dose: 5 to 10 minims (0.3 to 0.6 C.c.).

FŒNICULUM (Fennel).

The fruit of Fœniculum capillaceum (nat. ord. Umbelliferæ), growing in Asia and Europe.

It contains a volatile and a fixed oil.

It is aromatic.

OLEUM FŒNICULI (Oil of Fennel).

A volatile oil distilled from fennel, and contains Anethol $C_{10}H_{12}O$ and a hydrocarbon, $C_{10}H_{16}$.

It should be kept in well-stoppered bottles, in a cool place, and if it be partially or wholly solidified, should be completely liquefied by warming before being dispensed. Soluble in alcohol and glaciale acetic acid.

It is not colored by the addition of ferric chloride test solution (absence of foreign oils containing phenols and carbolic acid).

Official in the water, compound licorice powder and compound spirit juniper. Dose: 5 to 15 drops (0.3 to 0.9 C.c.).

AQUA FŒNICULI (Fennel Water).

Prepared by triturating oil of fennel with precipitated calcium phosphate, mixing with water, and filtering.

Used principally as a vehicle for other medicines.

LECTURE NO. 17.

OLEUM GAULTHERIÆ (Oil of Gaultheria) Oil of Wintergreen.

Oil of Teaberry, Oil of Partridge-berry. A volatile oil distilled from the leaves of Gaultheria procumbens (nat. ord. Ericaceæ).

It consists entirely of methyl salicylate, and is nearly identical with volatile oil of birch. 169 grains of the oil contain 152 grains of methyl salicylate, and are equivalent to 138 grains of salicylic acid. It gives a purple color in aqueous solution, with ferric salts.

It is the heaviest volatile oil, having a specific gravity of 1.175 to 1.185 at 15°C (59°F). Soluble in alcohol.

Largely adulterated with chloroform. It should be colorless, but as found in commerce is often of a reddish color.

Official in the spirit. It is used for flavoring purposes, and as a substitute for salicylic acid.

Dose: 10 to 30 minims (0.6 to 1.8 C.c.).

SPIRITUS GAULTHERIÆ (Spirit of Gaultheria).

A solution of the oil in alcohol, containing 5%.

Used principally as a flavoring agent. Dose: 10 to 20 minims (0.6 to 1.20 C.c.).

HEDEOMA (Pennyroyal).

The leaves and tops of Hedeoma pulegioides (nat. ord. Labiatæ), growing in all parts of the United States.

It contains a Volatile Oil upon which its virtues depend.

It is a stimulant aromatic.

OLEUM HEDEOMÆ (Oil of Pennyroyal).

A volatile oil distilled from pennyroyal, and should be kept in well-stoppered bottles, in a cool place, away from the light.

It contains a body having the composition $C_{10}H_{18}O$, (hedeomol) one having the composition $C_{10}H_{17}O$, and one having the composition $C_6H_{12}O$, formic and acetic acids.

Soluble in alcohol, carbon disulphide, and glaciale acetic acid. Used in flatulent colic and sick stomach. Dose: 2 to 10 drops (0.22 to 0.6 C.c.).

ILLICIUM (Star Anise).

The fruit of Illicium verum (nat. ord. Magnoliaceæ) growing in China.

It contains Volatile Oil, (consisting of Anethol $C_{10}H_{12}O$ and a number of terpenes).

Used principally as a source of the oil which is substituted for the oil of anise.

OLEUM JUNIPERI (Oil of Juniper).

A volatile oil distilled from the fruit of Juniperus communis (nat. ord. Coniferæ).

It consists of two hydrocarbons, principally pinene, $C_{10}H_{16}$. Soluble in alcohol and carbon disulphide. Sometimes adulterated with oil of turpentine, which may be detected by the low specific gravity of the mixture.

Holland Gin owes its peculiar flavor and diuretic power to the oil of Juniper.

It is stimulant, diuretic and carminative. Official in the spirit and compound spirit.

Dose: 5 to 15 minims (0.3 to 0.9 C.c.).

SPIRITUS JUNIPERI (Spirit of Juniper).

A solution of oil of juniper in alcohol, and contains 5%.

Used principally as an addition to diuretic infusions. Dose: 30 to 60 minims (1.9 to 3.7 C.c.).

SPIRITUS JUNIPERI COMPOSITUS (Compound Spirit of Juniper).

Contains oil of Juniper, oil of caraway, oil of fennel, alcohol and water.

Used for the same purposes as the spirit. Dose: 2 to 4 fluid drachms (7.5 to 15 C.c.).

OLEUM LAVANDULÆ FLORUM (Oil of Lavender Flowers).

A volatile oil distilled from the fresh flowers of Lavandula officinalis (nat. ord. Labiatæ), and should be kept in well-stoppered bottles in a cool place, away from the light. It consists of linalool, linalool acetate, linalool butyrate, and a geraniol acetate.

Soluble in alcohol in all proportions (distinction from oil of turpentine) and in three times its volume of a mixture of three volumes of alcohol and one volume of water (distinction and absence of oil of turpentine). Also soluble in glaciale acetic acid, and forms a turbid mixture with carbon disulphide.

Oil of Spike is obtained from the broad-leaved variety of Lavender growing wild in Europe (Lavandula spica), which is used by artists in varnishes.

Official in the spirit, compound tincture and aromatic spirit of ammonia.

SPIRITUS LAVANDULÆ (Spirit of Lavender).

Prepared by dissolving the oil in deodorized alcohol, and contains 5%.

Used chiefly as a perfume. It is stimulant and carminative. Dose: 30 to 60 minims (1.9 to 3.7 C.c.).

TINCTURA LAVANDULÆ COMPOSITA (Compound Tincture of Lavender), Compound Spirit of Lavender, Lavender Drops.

Contains oil of lavender, oil of rosemary, cassia cinnamon, cloves, nutmeg, red saunders, alcohol and water.

Used principally as an adjuvant and corrigent of other medicines. Dose: 30 to 60 drops (1.9 to 3.75 C.c.).

LIMONIS CORTEX (Lemon Peel).

The rind of the recent fruit of Citrus Limonum (nat. ord. Rutaceæ), obtained chiefly from the West Indies.

It contains a Volatile Oil, and Hesperidin, a bitter principle.

Official in the spirit.

Used principally as a flavoring agent.

OLEUM LIMONIS (Oil of Lemon).

A volatile oil obtained by expressing from fresh lemon peel, and should be kept in well-stoppered bottles, in a cool place, away from the light.

It contains Limonene $C_{10}H_{16}$, another terpene isomeric with limonene, a sesquiterpene $C_{15}H_{24}$, and about $7\frac{1}{2}\%$ Citrol $C_{10}H_{16}O$.

Soluble in alcohol, carbon disulphide, and glaciale acetic acid. When kept for some time it should not develop a terebinthinate odor or taste (absence of oil of turpentine, or other oils, consisting chiefly of pinene).

Often adulterated with alcohol, oil of turpentine and fixed oils. Official in the spirit and compound spirit of orange.

Used principally as a flavoring agent.

SPIRITUS LIMONIS (Spirit of Lemon).

Contains oil of lemon 5%, lemon peel, and deodorized alcohol. Used as a flavoring agent.

MACIS (Mace).

The arillode of the fruit of Myristica fragrans (nat. ord. Myristicaceæ) growing in the East Indies. It contains 7% to 9% of Volatile Oil (consisting of macene $C_{10}H_{16}$, pinene, fixed oil, gummy matter and ligneous fibre.

It is aromatic.

MATRICARIA (Matricaria), German Chamomile.

The flower heads of Matricaria Chamomilla (nat. ord. Compositæ), growing in the United States and Europe.

It contains a Volatile Oil (consisting of a terpene $C_{10}H_{16}$, a colorless oil $C_{10}H_{16}O$, and bitter extractive.

It is a mild tonic.

MELISSA (Melissa) Balm.

The leaves and tops of Melissa officinalis (nat. ord. Labiatæ), growing in the United States and Europe.

It contains a Volatile Oil, tannin, bitter extractive and gum.

Very rarely used.

MENTHA PIPERITA (Peppermint).

The leaves and tops of Mentha Piperita (nat. ord. Labiatæ), growing in the United States and Europe.

It contains a Volatile Oil upon which its virtues depend.

Official in the spirit. It is an aromatic stimulant.

OLEUM MENTHÆ PIPERITÆ (Oil of Peppermint).

A volatile oil distilled from peppermint, containing Menthol $C_{10}H_{19}OH$ and terpenes.

Soluble in alcohol, carbon disulphide and glaciale acetic acid. Sometimes adulterated with oil of erigeron, castor oil, oil of turpentine, oil of copaiba, alcohol, etc.

Official in the spirit, water and troche. It is a stimulant and carminative. Dose: 2 to 6 drops (0.12 to 0.36 C.c.).

SPIRITUS MENTHÆ PIPERITÆ (Spirit of Peppermint).

Prepared by dissolving 10% of the oil in alcohol, adding peppermint leaves to color it, and filtering.

Dose: 10 to 20 minims (0.6 to 1.25 C.c.).

TROCHISCI MENTHÆ PIPERITÆ (Troches of Peppermint).

Contain oil of peppermint, sugar and mucilage of tragacanth.

Used in gastric and intestinal pains.

AQUA MENTHÆ PIPERITÆ (Peppermint Water).

Prepared by triturating the oil with precipitated calcium phosphate, adding water and filtering.

MENTHA VIRIDIS (Spearmint).

The leaves and tops of Mentha viridis (nat. ord. Labiatæ), growing in the United States and Europe.

It contains a volatile oil upon which its virtues depend.

Official in the spirit. Its uses are the same as peppermint.

OLEUM MENTHÆ VIRIDIS (Oil of Spearmint).

A volatile oil distilled from spearmint, and contains a terpene $C_{10}H_{16}$, and a compound identical with carvol $C_{10}H_{14}O$.

Soluble in alcohol. Official in water and spirit.

AQUA MENTHÆ VIRIDIS (Spearmint Water).

Prepared by triturating the oil with precipitated calcium phosphate, adding water and filtering.

SPIRITUS MENTHÆ VIRIDIS (Spirit of Spearmint).

Prepared by dissolving 10% of the oil in alcohol, adding spearmint leaves to color it, adding water and filtering.

OLEUM MYRCIÆ (Oil of Bay).

A volatile oil distilled from the leaves of Myrcia Acris (nat. ord. Myrtacæ), growing in the West Indies.

It consists of two oils, one heavy and the other light, containing Eugenol, pinene, dipentene, diterpene (insoluble in alcohol). With an equal volume of alcohol, glaciale acetic acid, or carbon disulphide it yields slightly turbid solutions. If, on being shaken with boiling water and passed through a wetted filter, the clear filtrate should produce, with a drop of ferric chloride test solution, only a transient grayish green, and not a blue or violet color (absence of carbolic acid).

Official in Spiritus Myrcia.

SPIRITUS MYRCIÆ (Spirit of Myrcia) Bay Rum.

Contains oil of bay, oil of orange peel, oil of pimenta, alcohol and water.

Used chiefly as a refreshing perfume.

MYRISTICA (Nutmeg).

The seed of Myristica fragrans (nat. ord. Myristicaceæ), growing in Sumatra, Java, and other parts of the East Indies.

It contains a volatile oil, water, ash, fixed oil and fat; starch, crude fibre and albuminoids.

Official in aromatic powder, compound tincture of lavender and vinegar of opium. It is aromatic and narcotic.

OLEUM MYRISTICÆ (Oil of Nutmeg).

A volatile oil distilled from nutmeg, and contains Myristocol $C_{10}H_{14}O$, and a stearopten which it deposits on standing, called Myristin.

Soluble in alcohol, ether and glaciale acetic acid.

Used for the same purposes as nutmeg. Dose: 2 to 3 drops (0.12 to 0.18 C.c.). The expressed oil of nutmeg is called OIL OF MACE, OR NUTMEG BUTTER.

PIMENTA (Pimenta) Allspice. Jamaica Pepper.

The nearly ripe fruit of Pimenta officinalis (nat. ord. Myrtaceæ) growing in Mexico, West Indies and South America. It is called Allspice on account of the fruit having a fragrant odor, thought to resemble that of a mixture of cinnamon, clove and nutmeg.

It contains a volatile oil, a green fixed oil, a fatty substance, tannin, gum, resin, uncrystallizable sugar, coloring matter, malic and gallic acids, saline matters, moisture and lignin.

It is a warm aromatic. Dose: 10 to 40 grains (0.65 to 2.6 gm.)

OLEUM PIMENTÆ (Oil of Allspice).

A volatile oil distilled from pimenta, and consists like oil of clove, of two distinct oils, a light and heavy, which are analagous to light and heavy oil of cloves.

Used for the same purpose as the other stimulant aromatic oils.

Official in spirit of myrcia.

OLEUM PICIS LIQUIDÆ (Oil of Tar).

A volatile oil distilled from tar. Soluble in alcohol.

It has the same medicinal properties as tar and is less offensive to the taste. Dose: 1 to 5 minims (0.06 to 0.3 C.c.).

ROSA CENTIFOLIA (Pale Rose) Cabbage Rose Petals.

The petals of Rosa centifolia (nat. ord. Rosaceæ) growing all over the world, and should be collected when the flower is fully expanded, but has not begun to fall.

It contains a Volatile Oil, malic and tartaric acid tannin, fat, resin, sugar, coloring matter.

Used principally in the preparation of the oil and rose water, but when taken internally it is slightly laxative.

AQUA ROSÆ FORTIOR (Stronger Rose Water) Triple Rose Water.

Water saturated with volatile oil of rose petals obtained as a by-product in the manufacture of oil of rose, and should be kept in well-stoppered bottles away from the light.

Used in making rose water.

AQUA ROSÆ (Rose Water).

Prepared by mixing together equal volumes of stronger rose water and distilled water, immediately before use. Official in the ointment. Used principally in eye waters, etc.

OLEUM ROSÆ (Oil of Rose) Attar, Otto or Essence of Roses.

A volatile oil distilled from the fresh flowers of Rosa Damascena in Northern India, from Rosa Moschata in Persia, and from Rosa Centifolia in the North of European Turkey. It consists of two oils, one liquid which is oxygenated and called Rhodinol $C_{10}H_{18}O$, and the other a stearopten $C_{16}H_{34}$.

Slightly soluble in alcohol. Sometimes adulterated with oil of gingergrass, oil of geranium, other volatile oils, wax, spermaceti, etc. The volatile additions to oil of rose may be detected by not being concrete, while the fixed oils may be detected by the greasy stain they leave on paper when heated. Used principally as a perfume.

OLEUM ROSMARINI (Oil of Rosemary).

A volatile oil distilled from the leaves of Rosmarinus officinalis (nat. ord. Labiatæ) and contains a hydrocarbon $C_{10}H_{16}$ called Pinene, another portion having a higher boiling point, which at a low temperature deposits a stearopten. Soluble in alcohol and glaciale acetic acid. Sometimes adulterated with oil of turpentine.

Official in compound tincture of lavender. It is stimulant, but is used chiefly as an ingredient in rubifacient liniments. Dose: 3 to 6 drops (0.18 to 0.36 C.c.).

SABINA (Savine) Savine Tops.

The tops of Juniperous Sabina (nat. ord. Coniferæ) growing in the United States, in the South of Europe and the Levant.

It contains a Volatile Oil, gum, tannic or gallic acids, resin, chlorophyll, fixed oil, bitter extractive, lime and salts of potassa.

The tops of Juniperus Virginiana or Common Red Cedar are sometimes substituted for savine, as they closely resemble each other.

Official in the fluid extract. It is highly irritant.

EXTRACTUM SABINÆ FLUIDUM (Fluid Extract of Savine).

Prepared from savine No. 40 in the usual way, using alcohol as the menstruum.

Rarely given internally. Dose: 3 to 8 minims (0.18 to 0.5 C.c.).

OLEUM SABINÆ (Oil of Savine).

A volatile oil distilled from savine, consisting principally of an oxygenated portion $C_{10}H_{16}O$ (Tilden). Soluble in alcohol, and glaciale acetic acid.

It is a powerful local irritant. Rarely used. Dose. 2 to 5 drops (0.12 to 0.3 C.c.).

OLEUM SANTALI (Oil of Santal).

A volatile oil distilled from the wood of Santalum Album (nat. ord. Santalaceæ) growing in India, and consists of Santalol $C_{15}H_{26}O$ and Santalal $C_{15}H_{24}O$.

Often adulterated with castor oil, other fixed oils and volatile oil of cedar. Soluble in alcohol.

Used chiefly as a perfume, also in rheumatism, gout, etc.

Dose: 15 to 20 minims (0.9 to 1.25 C.c.).

SALVIA (Sage).

The leaves of Salvia officinalis (nat. ord. Labiatæ) growing in the United States and Europe.

It contains a Volatile Oil, upon which its virtues depend (consisting of two terpenes having different boiling points, Salviol $C_{10}H_{16}O$ and camphor $C_{10}H_{16}O$). In the fresh oil the terpene having a lower boiling point predominates, but on standing the salviol increases and then the camphor.

It is tonic, astringent and aromatic. Dose: 20 to 30 grains (1.3 to 1.95 gm.).

SAMBUCUS (Elder).

The flowers of Sambucus Canadensis (nat. ord. Caprifoliaceæ) growing in the United States. It contains a Volatile Oil (consisting of a hydrocarbon, Sambucene $C_{10}H_{16}$ and probably a stearopten).

They are excitant and sudorific, but are seldom used.

SASSAFRAS (Sassafras).

The bark of the root of sassafras varifolium (nat. ord. Laurineæ) growing in the United States.

It contains a Volatile Oil, camphoraceous matter, fatty matter, resin, wax, a peculiar decomposition product of tannic acid called sassafrid, tannic acid, gum, albumen, starch, lignin and salts.

It is an aromatic stimulant and astringent.

OLEUM SASSAFRAS (Oil of Sassafras).

A volatile oil distilled from Sassafras, consisting of a hydrocarbon Safrene $C_{10}H_{16}$, and an oxidized compound, safrol $C_{10}H_{10}O_2$. Soluble in alcohol, carbon disulphide and glaciale acetic acid. It becomes darker and thicker by age and exposure to air.

Used chiefly for flavoring purposes.

OLEUM SINAPIS VOLATILE (Volatile Oil of Mustard).

A volatile oil obtained from Black Mustard, by maceration with water and subsequent distillation after previously expressing the fixed oil.

It consists of Allyl Isosulphocyanate C_3H_5CNS. Often produced artificially by the action of an alcoholic solution of potassium sulphocyanate on allyl iodide. The oil does not pre-exist in the plant. Often adulterated with other oils, which may be detected by mixing 50 drops of concentrated sulphuric acid with 5 drops of the suspected oil in a glass tube. If the oil is pure, little change of color will be produced, but if adulterated, a red or brown color will soon appear.

Soluble in alcohol, ether and carbon disulphide. Official in compound liniment of mustard.

LINIMENTUM SINAPIS COMPOSITUS (Compound Liniment of Mustard).

Contains volatile oil of mustard, fluid extract of mezereum, camphor, castor oil and alcohol.

The ext. of mezereum is a very energetic irritant, and the castor oil is used to increase the consistency of fhe liniment. It is irritant.

SERPENTARIA (Serpentaria) Virginia Snakeroot.

The rhizome and rootlets of Aristolochia Serpentaria, and Aristolochia reticulata (nat. ord. Aristolochiaceæ) growing in the Middle, Southern and Western States.

It contains Volatile Oil (consisting of a terpene $C_{10}H_{16}$, a compound ester, a crystalline acid, and a small portion having the formula $C_{18}H_{30}O$), yellowish green resin, extractive matter, gummy extract, lignin and water.

Official in the fluid extract, tincture, and compound tincture of cinchona.

It is a stimulant tonic, diuretic and diaphoretic.

EXTRACTUM SERPENTARIÆ FLUIDUM (Fluid Extract of Serpentaria).

Prepared in the usual manner from serpentaria No. 60, using a mixture of 800 parts of alcohol and 200 parts of water as the menstruum.

Dose: 20 to 30 minims (1.25 to 1.9 C.c.).

TINCTURA SERPENTARIÆ (Tincture of Serpentaria).

Prepared by macerating and percolating serpentaria No. 40 with a mixture of 650 parts of alcohol and 350 parts of water.

Dose: 1 to 4 fluid drachms (3.75 to 15 C.c.).

TANACETUM (Tansy).

The leaves and tops of Tanacetum vulgare (nat ord. Compositæ) growing in the United States.

It contains a Volatile Oil (consisting of a turpene $C_{10}H_{16}$, an aldehyde $C_{10}H_{16}O$, and an alcohol $C_{10}H_{18}O$) tanacetin $C_{11}H_{16}O_4$, tannic acid, traces of gallic acid, a wax-like substance, albuminoids, tartaric, citric and malic acids, traces of oxalic acid, a sugar, resin, metarabic acid, pararabin, and woody fibre. It is aromatic and irritant narcotic.

Dose: 30 to 60 grains (1.95 to 3.9 gm.).

LECTURE NO. 18.

OLEUM TEREBINTHINÆ (Oil of Turpentine) Spirit of Turpentine.

A volatile oil distilled from turpentine.

As found in commerce it contains oxygen, but when pure contains only hydrogen and carbon, and consists of one or more turpenes. Very volatile and inflammable, and less soluble in alcohol than most volatile oils. Soluble in ether. If a little of the oil be evaporated in a capsule on a water bath, it should have not more than a slight residue (absence of petrolatum, paraffine oils and resin).

Official in the liniment and is used in preparing the rectified oil, terebene and terpin hydrate. It is stimulant, diuretic, diaphoretic and anthelmintic. Dose: 5 to 30 drops (0.3 to 1.9 C.c.) Externally it is rubifacient.

OLEUM TEREBINTHINÆ RECTIFICATUS (Rectified Oil of Turpentine).

Prepared by distilling oil of turpentine with lime water and collecting the product, and should be kept in well-stoppered bottles, in a cool place, away from light. Should be dispensed when oil of turpentine is prescribed for internal use.

Distilling with lime frees it from the products which give the commercial oil its disagreeable odor and taste.

LINIMENTUM TEREBINTHINÆ (Turpentine Liniment).

Contains resin cerate dissolved in oil of turpentine. Used as a remedy in burns and scalds.

TEREBENUM (Terebene) $C_{10}H_{16}$.

A liquid consisting chiefly of pinene, and containing not more than a very small proportion of terpene and dipentine, and should be kept in well-stoppered bottles, in a cool place, away from the light.

Prepared by the action of sulphuric acid on oil of turpentine.

Slightly soluble in water, but soluble in an equal bulk of alcohol, glaciale acetic acid and carbon disulphide.

It is a valuable stimulant expectorant. Dose: 20 to 60 minims (1.2 to 3.7 C.c.), given in emulsion or capsules.

TERPINI HYDRAS (Terpin Hydrate) $C_{10}H_{18}(OH)_2$. H_2O.

A hydrate of the diatomic alcohol Terpin, and should be kept in well-stoppered bottles.

Prepared by mixing together oil of turpentine, alcohol and nitric acid; on standing, the terpin hydrate crystallizes out.

Soluble in alcohol, and sparingly soluble in water.

Used in chronic bronchitis. Dose: 2 to 3 grains (0.13 to 0.2 gm.), given in pill or emulsion.

OLEUM THYMI (Oil of Thyme).

A volatile oil distilled from the leaves and flowering tops of Thymus vulgaris (nat. ord. Labiatæ) growing in France and cultivated in our gardens.

Soluble in alcohol, carbon disulphide and glaciale acetic acid, and contains cymene $C_{10}H_{14}$, thymene $C_{10}H_{16}$ and Thymol $C_{10}H_{14}O$.

It become darker and thicker by age and exposure to air. The impure oil is called OIL OF ORIGANUM.

Dose : 3 to 15 drops (0.3 to 0.9 C.c.).

VALERIANA (Valerian).

The rhizome and roots of Valeriana officinalis (nat. ord. Valerianeæ) growing in Europe.

It contains Volatile Oil (consisting of a hydrocarbon Valerene $C_{10}H_{16}$ and Valerol $C_6H_{10}O$, which slowly oxidizes into valerianic acid) a peculiar extractive matter, gum, a soft odorous resin and lignin.

Official in the tincture, fluid extract and ammoniated tincture. It is a nerve stimulant.

EXTRACTUM VALERIANÆ FLUIDUM (Fluid Extract of Valerian).

Prepared by the usual method from valerian No. 60, using a mixture of 750 parts of alcohol and 250 parts of water as the menstruum.

Dose : 1 fluid drachm (3.75 C.c.).

TINCTURA VALERIANÆ (Tincture of Valerian).

Prepared by macerating and percolating valerian No. 60 with a mixture of 750 parts of alcohol and 250 parts of water.

It deposits on standing a black, very cohesive precipitate with starch and a yellow extractive matter. Dose : 1 to 4 fluid drachms(3.75 to 15C.c.).

TINCTURA VALERIANÆ AMMONIATA (Ammoniated Tincture of Valerian).

Prepared by macerating and percolating valerian No. 60 with aromatic spirit of ammonia.. Dose : 30 to 60 minims (1.9 to 3.75 C.c.).

STEAROPTENS.

CAMPHORA (Camphor) $C_{10}H_{16}O$.

A stearopten obtained from Cinnamomum Camphora (nat. ord. Laurineæ) (having the nature of a ketone) by cutting the roots, trunks and branches in small chips, heating with water and condensing the camphor on rice straw, then purifying by sublimation.

Camphor in the crude state is brought into this country from Canton, Batavia, Singapore and Calcutta.

It is easily powdered by the aid of a little alcohol, ether or chloroform. On exposure to the air it evaporates, and when heated sublimes without leaving any residue. When triturated with menthol, thymol, phenol or chloral hydrate liquefaction takes place.

Sparingly soluble in water, but soluble in alcohol, ether, chloroform, carbon disulphide, benzin, and fixed oils. By the intervention of magnesia or sugar, a larger quantity may be dissolved in water.

It is refined by melting with lime and then subliming; the lime removing the water which interferes with the solidification of the vapor. Sometimes adulterated with artificial camphor. Camphor $C_{10}H_{16}O$ and Borneol $C_{10}H_{18}O$ (Sumatra Camphor, Borneo Camphor obtained from Sumatra and Borneo) are classified together as belonging to the group called in general camphors, which occur with the terpenes or essential oils $C_{10}H_{16}$, and are to be considered as oxidation products of these latter.

Borneol is a secondary alcohol and therefore contains the group CHOH linked to a more complex group. Ketones are formed from secondary alcohols by oxidation and the group CHOH changed to CO. Camphor bears this relation to borneol, and is therefore considered as a ketone, although not capable of being formed directly from borneol.

Official in the water, liniment, spirit, soap liniment, camphorated tincture of opium.

It is stimulant and diaphoretic. Dose: 5 to 10 grains (0.33 to 0.65 gm.).

AQUA CAMPHORÆ (Camphor Water).

Prepared by powdering camphor with the aid of a little alcohol, mixing with precipitated calcium phosphate, adding water and filtering. It contains 0.8% of camphor. Dose: 1 to 2 tablespoonfuls (15 to 30 C.c.).

LINIMENTUM CAMPHORÆ (Camphor Liniment).

Contains 20% of camphor dissolved in cotton-seed oil.

Used in cases of sprains and bruises.

SPIRITUS CAMPHORÆ (Spirit of Camphor).

Contains 10% of camphor dissolved in alcohol.

It is an anodyne embrocation. The spirit of the U. S. P. 1880 contained water.

CAMPHORA MONOBROMATA (Monobromated Camphor) Brominated Camphor. Bromated Camphor.

Prepared by heating camphor and bromine together in a sealed tube.

Insoluble in water, but soluble in alcohol, ether, chloroform, fixed and volatile oils.

Used as a sedative. Dose: 5 grains (0.33 gm.).

MENTHOL (Menthol) $C_{10}H_{19}OH$.

A stearopten having the character of a secondary alcohol obtained from the oil of peppermint by cooling. It should be kept in well-stoppered bottles, in a cool place. Soluble in alcohol, ether, chloroform, carbon disulphide and glaciale acetic acid. When triturated with an equal weight of camphor, thymol, or chloral hydrate, the mixture becomes liquid. It is often adulterated with spermaceti, paraffine, etc.

Used as a local anæsthetic application.

THYMOL. See page 105.

Glucosides and Neutral Principles.

ALOE BARBADENSIS (Barbadoes Aloes).

The inspissated juice of the leaves of Aloe vera (nat. ord. Liliaceæ) growing in the East Indies and Barbadoes.

Used considerably to adulterate Socotrine Aloes.

ALOE SOCOTRINA (Socotrine Aloes).

The inspissated juice of the leaves of Aloe Perryi (nat. ord. Liliaceæ) growing iu Socotra.

There are three varieties of aloes in commerce, namely: Barbadoes, Socotrine and Cape. Cape Aloes was dropped by the U. S. P. 1880. It comes from the Cape of Good Hope and differs from the Socotrine in its brilliant conchoidal fracture and peculiar strong odor. Socotrine Aloes is found in pieces of a yellowish or reddish brown color, sometimes very light, especially when fresh, as it becomes darker by exposure to the air and is the best variety. There is stlll another variety called Natal Aloes, coming from Natal on the South East coast of Africa. Its odor is like that of Cape Aloes but it is less soluble. Barbadoes Aloes is not uniform in color, sometimes being dark brown or almost black, sometimes of a reddish brown or liver color, or orange brown. It is distinguished by its odor, which is disagreeable and nauseous.

The principle upon which the action of aloes depends is Aloin. Aloes yields its active matter to cold water, and when good is almost wholly dissolved by boiling water. Soluble in alcohol. It is inflammable, giving off a thick smoke which has the odor of the drug.

Official in the extract and is used in preparing the purified aloes. It is cathartic. Dose: 10 grains (0.65 gm.), best administered in pills on account of its bitter taste.

EXTRACTUM ALOES (Extract of Aloes).

Prepared by mixing Socotrine aloes with boiling water, letting macerate, pouring off the clear liquid and evaporating to dryness.

Dose: 2 to 10 grains (0.13 to 0.65 gm.).

ALOE PURIFICATA (Purified Aloes).

Prepared by heating Socotrine Aloes on a water bath until it is completely melted, adding alcohol, straining, evaporating with constant stirring until a thread of the mass becomes brittle on cooling, breaking into small pieces and keeping in well-stoppered bottles.

The impurities that are removed by this process are fragments of wood, vegetable remains, pieces of leather and earthy matter. Almost entirely soluble in alcohol.

Official in compound extract of colocynth, pill, pill of aloes and asafetida, pill of aloes and iron, pill of aloes and mastic, pill of aloes and myrrh, compound rhubarb pills, tincture, tincture of aloes and myrrh, compound tincture of benzoin.

PILULÆ ALOES (Pills of Aloes).

Contain purified aloes, soap and water.

Each pill contains 2 grains of aloes. Dose: 5 pills.

PILULÆ ALOES ET ASAFŒTIDÆ (Pills of Aloes and Asafetida).

Contain purified aloes, asafetida, soap and water.

Each pill contains about 4 grains of the mass. Dose: 2 to 5 pills.

PILULÆ ALOES ET FERRI (Pills of Aloes and Iron).

Contain purified aloes, dried ferrous sulphate, aromatic powder and confection of rose. Dose: 1 to 3 pills.

PILULÆ ALOES ET MASTICHES (Pills of Aloes and Mastic). Lady Webster's Dinner Pills.

Contain purified aloes, mastic, red rose and water.

Each pill contains about 3 grains of the solid ingredients, and nearly 2 grains of aloes.

PILULÆ ALOES ET MYRRHÆ (Pills of Aloes and Myrrh) Rufus's Pills.

Contain purified aloes, myrrh, aromatic powder and syrup. Dose: 3 to 6 pills, or from 10 to 20 grains (0.65 to 1.3 gm.) of the mass.

TINCTURA ALOES (Tincture of Aloes).

Prepared by macerating and percolating a mixture of purified aloes and licorice root with diluted alcohol.

Dose as a purgative: 2 to 4 fluid drachms (7.5 to 15 C.c.); as a laxative: ½ to 1 fluid drachm (1.9 to 3 7 C.c.).

TINCTURA ALOES ET MYRRHÆ (Tincture of Aloes and Myrrh). Elixir Proprietatis, Compound Tincture of Aloes.

Prepared by macerating and percolating a mixture of purified aloes, myrrh and licorice root with a mixture of three parts alcohol and one part water.

Dose: 1 to 2 fluid drachms (3.7 to 7.5 C.c.).

ALOINUM (Aloin) $C_{16}H_{18}O_7$.

A neutral principle obtained from several species of Aloes. Barbadoes aloes yields Barbaloin $C_{17}H_{20}O_7$ Socotrine aloes, Socaloin $C_{17}H_{10}O_7$, and Natal aloes, Nataloin $C_{16}H_{16}O_7$. These aloins differ more or less in composition and properties, according to their source. Aloin is soluble in water and alcohol.

They are distinguished from one another as follows: To aloes or powdered aloes, on a white plate add strong nitric acid: Socaloin produces no color, while Nataloin and Barbaloin produce a crimson color. To another portion add strong sulphuric acid and the vapor of nitric acid: Nataloin produces a blue color, while Barbaloin does not.

Aloin is rapidly decomposed in alkaline solution. It is an active purgative. Dose: ½ to 2 grains (0.032 to 0.13 gm.).

AMYGDALA AMARA (Bitter Almond).

The seed of Prunus Amygdalus, var. amara, (nat. ord. Rosaceæ) growing in Persia, Barbary and Syria.

It contains a glucoside, Amygdalin and a ferment emulsin.

When the almonds are exposed to moisture, fermentation is produced and the bitter oil and hydrocyanic acid are formed, but these do not preexist in the plant. The changes that take place are as follows:

$C_{20}H_{27}NO_{11}$ (amygdalin) $+$ 2 H_2O (water) $=$ C_7H_5OH (oil of bitter almonds) $+$ HCN (hydrocyanic acid) $+$ 2 $C_6H_{12}O_6$ (glucose).

It is sedative. Official in the syrup.

APOCYNUM (Canadian Hemp).

The root of Apocynum cannabium (nat. ord. Apocynaceæ) growing principally in the Western part of the United States.

It contains a glucoside, Apocynein, soluble in water and an amorphous resinous substance Apocynin, soluble in alcohol and ether, but almost insoluble in water.

Apocynum is sometimes improperly called Indian Hemp. Official in the fluid extract. It is emetic, cathartic and diuretic.

EXTRACTUM APOCYNI FLUIDUM (Fluid Extract of Apocynum).

Prepared in the usual manner from apocynum No. 60, using a mixture of alcohol, water and glycerin as the menstruum.

Dose: 5 minims (0.3 C.c.).

ASCLEPIAS (Pleurisy Root) Butterfly Weed.

The root of Asclepias tuberosa (nat. ord. Asclepideæ) growing throughout the United States.

It contains a crystalline glucoside, tannic and gallic acids, albumen, pectin, gum, starch, a resin soluble and another insoluble in ether, fixed oil, volatile odorous fatty matter, various salts and lignin.

Official in the fluid extract. It is diaphoretic and expectorant. Dose: 20 to 60 grains (1.3 to 3.9 gm.).

EXTRACTUM ASCLEPIADIS FLUIDUM (Fluid Extract of Asclepias).

Prepared in the usual manner from asclepias No 60, using dilute alcohol as the menstruum.

Dose: 20 to 60 minims (1.25 to 3.7 C.c.).

BRYONIA (Bryony).

The root of Bryonia alba and Bryonia dioica (nat. ord. Cucurbitaceæ) growing in different parts of Europe. Containing a glucoside Bryonin, $C_{48}H_{80}O_{19}$, starch, gum, resin, sugar, a concrete oil, albumen and various salts.

Official in the tincture. It is a hydragogue cathartic. Dose: 20 to 60 grains (1.3 to 3.9 gm.).

TINCTURA BRYONIÆ (Tincture of Bryonia).

Prepared by macerating and percolating recently dried bryonia No. 40 with alcohol.

Dose: 1 to 2 fluid drachms (3.7 to 7.5 C.c.).

CALENDULA (Marigold).

The florets of Calendula officinalis (nat. ord. Compositæ) growing in all parts of the United States.

Contains a peculiar principle Calendulin, upon which its virtues depend.

Official in the tincture.

TINCTURA CALENDULÆ (Tincture of Calendula).

Prepared by macerating and percolating calendula No. 20 with alcohol. Used externally for the same purposes as arnica flowers.

CASCARILLA (Cascarilla) Cascarilla Bark.

The bark of Croton Elutaria (nat. ord. Euphorbiaceæ) growing in the Bahama Islands.

Contains a bitter crystallizable principle called Cascarillin, albumen, a peculiar kind of tannin, a red coloring matter, fatty matter of a nauseous odor, wax, gum, volatile oil, resin, starch, pectic acid, potassium chloride, a salt of lime and lignin.

It is aromatic and tonic.

CAULOPHYLLUM (Blue Cohosh) Pappoose Root, Squaw Root, Blueberry Root.

The rhizome and roots of Caulophyllum thalictroides (nat. ord. Berberidaceæ) growing in the United States.

Contains a glucoside called Leontin (Lloyd), albumen, gum, starch, phosphoric acid, extractive, two resins, coloring matter.

It is sedative, antispasmodic and oxytoxic.

CHIMAPHILA (Pipsissewa) Prince's Pine, Wintergreen.

The leaves of Chimaphila umbellata (nat. ord. Ericaceæ) growing in all parts of the United States.

Contains a peculiar substance called Chimaphilin, gum, starch, sugar, extractive, pectic acid, tannic acid, resin, fatty matter, chlorophyll, yellow coloring matter, lignin and various inorganic substances.

Official in the fluid extract. It is diuretic, tonic and astringent.

EXTRACTUM CHIMAPHILÆ FLUIDUM (Fluid Extract of Chimaphila).

Prepared in the usual manner from chimaphila No. 30, using dilute alcohol as the menstruum. Dose: 1 fluid drachm (3.75 C.c.).

CHIRATA (Chirata) Chiretta.

The entire plant of Swertia Chirata (nat. ord. Gentianeæ) growing in Northern India.

Contains a peculiar bitter principle called Chiratin, resin, a peculiar bitter substance, brown coloring matter, gum, sugar, wax, chlorophyll, soft resin, tannin and ophelic acid.

Official in the fluid extract and tincture. It is a tonic.

EXTRACTUM CHIRATÆ FLUIDUM (Fluid Extract of Chirata).

Prepared in the usual way from chirata No. 30, using a mixture of 2 parts of alcohol and 1 part water as the menstruum.

Dose : ½ fluid drachm (1.9 C.c.).

TINCTURA CHIRATÆ (Tincture of Chirata).

Prepared by macerating and percolating chirata No. 40, with a mixture of 650 parts alcohol and 350 parts water.

Dose : 1 to 2 fluid drachms (3.7 to 7.5 C.c.).

CHRYSAROBINUM (Crysarobin) Goa Powder, Araroba.

A neutral principle obtained from Goa Powder, a substance deposited in the wood of Andira Araroba (nat. ord. Leguminosæ) growing in the East Indies, the oldest trees yielding the largest amount of powder.

Contains resin, woody fibre, bitter extractive, and a substance called by Attfield, Chrysophanic Acid. It is slightly soluble in water and alcohol, but soluble in boiling benzol, and solutions of alkalies. It turns brownish-yellow when exposed to the air.

When taken in sufficient doses it is a gastro-intestinal irritant, but is used principally externally in skin diseases. Official in the ointment.

UNGUENTUM CHRYSAROBINI (Ointment of Chrysarobin).

Prepared by rubbing chrysarobin with benzoinated lard, and contains 5%.

Used in skin diseases.

COLOCYNTHIS (Colocynth).

The fruit of Citrullus Colocynthis (nat. ord. Cucurbitaceæ) deprived of its rind, growing in Turkey, Asia and Africa.

Contains a glucoside Colocynthin, extractive, fixed oil, a resinous substance insoluble in ether, gum, pectic acid, gummy extract, etc.

Official in the extract.

It is a powerful drastic hydragogue cathartic. Dose : 5 to 10 grains (0.33 to 0.65 gm.).

EXTRACTUM COLOCYNTHIDIS (Extract of Colocynth).

Prepared by macerating 1000 grams of ground or bruised colocynth, dried and freed from the seeds, with dilute alcohol for four days, expressing, percolating the residue with dilute alcohol until the tincture and expressed liquid measure 500 C.c., distilling off the alcohol on a water bath, evaporating to dryness and powdering.

It should be kept in well-stoppered bottles. Official in the compound extract.

LECTURE NO. 19.

EXTRACTUM COLOCYNTHIDIS COMPOSITUM (Compound Extract of Colocynth).

Contains extract of colocynth, purified aloes, cardamom No. 60, resin of scammony, soap and alcohol, and should be kept in well-stoppered bottles.

Official in compound cathartic pill and vegetable cathartic pill. It is cathartic. Dose: 5 to 30 grains (0.33 to 1.95 gm.).

PILULÆ CATHARTICÆ VEGETABILES (Vegetable Cathartic Pill).

Contains compound extract of colocynth, extract of hyoscyamus, extract of jalap, extract of leptandra, resin of podophyllum, oil of peppermint and water.

The oil of peppermint and extract of hyoscyamus are added to prevent griping. Dose: 3 pills.

CONVALLARIA (Lily of the Valley).

The rhizome and roots of Convallaria majalis (nat. ord. Liliaceæ), growing in the United States and Europe.

Contains two glucosides Convallarin $C_{34}H_{62}O_{11}$, and Convallamarin $C_{23}H_{44}O_{12}$. Official in the fluid extract. It is a cardiac stimulant.

EXTRACTUM CONVALLARIÆ FLUIDUM (Fluid Extract of Convallaria).

Prepared in the usual manner from convallaria No. 60, using dilute alcohol as the menstruum.

Dose: 5 to 15 minims (0.3 to 0.9 C.c.).

CROCUS (Saffron).

The stigmas of Crocus sativus (nat. ord. Irideæ) growing in Greece and Asia Minor.

Contains a glucoside Polychroite, odorous volatile oil, wax, gum, albumen, saline matter, water and lignin. Polychroite may be decomposed into crocin a coloring principle. According to M. Henry the value of crocus really depends on its volatile oil. Often adulterated with safflower, marigold, arnica, fibre of dried beef, red saunders, etc. Official in the tincture.

It should be kept in well-stoppered bottles. It is antispasmodic. Dose: 10 to 30 grains (0.65 to 1.95 gm.).

TINCTURA CROCI (Tincture of Crocus).

Prepared by macerating and percolating crocus with dilute alcohol.

Dose: 1 to 3 fluid drachms (3.7 to 11.25 C.c.).

CYPRIPEDIUM (Ladies' Slipper).

The rhizome and roots of Cypripedium pubescens, and Cypripedium parviflorum (nat. ord. Orchideæ) growing in the United States.

Contains a volatile oil, volatile acid, tannic and gallic acids, two resins, gum, glucose, starch and lignin. The virtues of the root probably reside in the volatile oil and a bitter principle.

Official in the fluid extract. It is stimulant and antispasmodic.

EXTRACTUM CYPRIPEDII FLUIDUM (Fluid Extract of Cypripedium).

Prepared in the usual manner from cypripedium No. 60, using dilute alcohol as the menstruum.

Dose: 15 minims (0.9 C.c.).

DIGITALIS (Foxglove).

The leaves of Digitalis purpurea (nat. ord. Scrophularineæ) growing in Europe, collected from plants of the second year's growth.

Contains a glucoside Digitalin, digitonin, digitalein, digitoxin, volatile oil, fatty matter, a red coloring substance, chlorophyll, albumen, starch, sugar, gum, lignin, etc.

Official in the infusion, extract, fluid extract and tincture.

It is a cardiac stimulant. Dose: 1 grain (0.65 gm.) When applied externally it acts as a diuretic.

INFUSUM DIGITALIS (Infusion of Digitalis).

Prepared by pouring boiling water over digitalis, bruised, macerating, straining and adding cinnamon water and water.

Dose: ½ fluid ounce (15 C.c.).

EXTRACTUM DIGITALIS (Extract of Digitalis).

Prepared by percolating digitalis No. 60 with a mixture of 2 parts alcohol and 1 part water, to exhaustion and evaporating to a pilular consistence.

Dose: ¼ grain (0.016 gm.).

EXTRACTUM DIGITALIS FLUIDUM (Fluid Extract of Digitalis).

Prepared in the usual manner from digitalis No. 60, using a mixture of 2 parts alcohol and 1 part water as the menstruum.

Dose: 1 or 2 minims (0.6 to 0.12 C.c.).

TINCTURA DIGITALIS (Tincture of Digitalis).

Prepared by macerating and percolating digitalis No. 60 with dilute alcohol.

Dose: 10 to 20 drops (0.6 to 1.25 C.c.).

ELATERINUM (Elaterin) $C_{20}H_{28}O_5$.

A neutral principle obtained from Elaterium (nat. ord. Cucurbitaceæ) a substance found deposited in the juice of the squirting cucumber, prepared by exhausting elaterium with chloroform, precipitating with ether, washing with ether, recrystallizing from chloroform.

Sparingly soluble in water, but soluble in alcohol. Official in the trituration.

It is a powerful hydragogue cathartic. Dose: $\frac{1}{20}$ grain (0.003 gm.).

TRITURATIO ELATERINI (Trituration of Elaterin).
Prepared by triturating elaterin with sugar of milk, and contains 10%.
Dose : ½ to ⅓ of a grain (0.03 to 0.04 gm.).

EUONYMUS (Wahoo).
The bark of the root of Euonymus atropurpurens (nat. ord. Celastrineæ) growing in the United States.
Contains a bitter principle called Euonymin, asparagin, a soft resin, a crystalline resin, a yellow resin, a brown resin, fixed oil, wax, starch, albumen, glucose, pectin, etc. Official in the extract.
It is a mild cathartic.

EXTRACTUM EUONYMI (Extract of Euonymus).
Prepared by exhausting euonymus No. 30 with a mixture of 2 parts alcohol and 1 part water, and evaporating to a pilular consistence.
Dose : 1 to 3 grains (0.065 to 0.2 gm.).

EUPATORIUM (Thoroughwort) Boneset, Indian Sage.
The leaves and flowering tops of Eupatorium perfoliatum (nat. ord. Compositæ) growing in the United States.
Contains a glucoside Eupatorin, chlorophyll, resin, crystalline matter, gum, tannin, yellow coloring matter, extractive, etc.
Official in the fluid extract. It is tonic, diaphoretic and in large doses emetic.

EXTRACTUM EUPATORII FLUIDUM (Fluid Extract of Eupatorium).
Prepared in the usual manner from eupatorium No. 40, using dilute alcohol as the menstruum. Dose : 20 to 60 minims (1.25 to 3.75 C.c.).

FRANGULA (Buckthorn) Alder Buckthorn.
The bark of Rhamnus Frangula (nat. ord. Rhamneæ) collected at least one year before being used. It grows in Europe and Russian Asia.
Contains a glucoside Frangulin, upon which its properties depend.
Official in the fluid extract. It is cathartic.

EXTRACTUM FRANGULÆ FLUIDUM (Fluid Extract of Frangula).
Prepared in the usual manner from frangula No. 40, using a mixture of 500 parts alcohol and 800 parts water as the menstruum.
Dose: 10 to 20 minims (0.6 to 1.25 C.c.).

GENTIANA (Gentian).
The root of Gentiana lutea (nat. ord. Gentianæ) growing in Europe.
Contains a neutral principle Gentiopicrin, gentisin, resinous matter, etc. Official in the extract, fluid extract and compound tincture. It is a tonic.

EXTRACTUM GENTIANÆ (Extract of Gentian).
Prepared by macerating and percolating gentian No. 20 with water to exhaustion and evaporating to a syrupy consistence.
Dose : 5 to 10 grains (0.33 to 0.65 gm.).

EXTRACTUM GENTIANÆ FLUIDUM (Fluid Extract of Gentian).

Prepared in the usual manner from gentian No. 30, using dilute alcohol as the menstruum.

Dose: 10 to 30 minims (0.6 to 1.9 C.c.).

TINCTURA GENTIANÆ COMPOSITA (Compound Tincture of Gentian).

Prepared by macerating and percolating a mixture of gentian No. 40, bitter orange peel No. 40 cardamom No. 40 with a mixture of 600 parts alcohol and 400 parts water.

Dose: 1 to 2 fluid drachms (3.7 to 7.5 C.c.).

GLYCYRRHIZA (Licorice).

The root of Glycyrrhiza glabra, var. grandulifera (nat. ord. Leguminoseæ) growing in Europe, Persia, Barbary and England.

Contains a glucoside, Glycyrrhizin $C_{24}H_{36}O_9$, a crystalline principle agedoite, starch, albumen, a brown acrid resin, a brown nitrogenous extractive matter, lignin, etc. It is claimed by Roussin that the sweet taste of the root is due to a compound ammonia with the glycyrrhizin and not to the glycyrrhizin alone.

Official in the fluid extract, extract, pure extract, and compound licorice powder. It is an excellent demulcent.

EXTRACTUM GLYCRRHIZÆ FLUIDUM (Fluid Extract of Licorice).

Prepared in the usual manner from licorice No. 40, using a mixture of alcohol, water and ammonia water as the menstruum.

The object of the ammonia water is to render the glycyrrhizin soluble.

EXTRACTUM GLYCYRRHIZÆ PURUM (Pure Extract of Licorice).

Prepared by exhausting licorice with a mixture of water and ammonia water, and evaporating to a pilular consistence.

Official in compound licorice mixture.

MISTURA GLYCYRRHIZÆ COMPOSITA (Compound Licorice Mixture). Brown Mixture.

Contains pure extract of licorice, syrup, mucilage of acacia, camphorated tincture of opium, wine of antimony, spirit of nitrous ether and water.

Dose: 1 to 2 tablespoonfuls (15 to 30 C.c.).

EXTRACTUM GLYCYRRHIZÆ (Extract of Licorice).

Prepared by boiling licorice with water, and evaporating to the proper consistence.

Official in troches of ammonium chloride, and troches of licorice and opium, and pill iodide of iron.

PULVIS GLYCYRRHIZÆ COMPOSITUS (Compound Licorice Powder).

Contains Senna No. 80, licorice No. 80, washed sulphur, sugar and oil of fennel. Dose: 30 to 60 grains (1.9 to 3.9 gm.).

GLYCYRRHIZINUM AMMONIATUM (Ammoniated Glycyrrhizin).

Prepared by percolating licorice with a mixture of water and ammonia water, precipitating the glycyrrhizin with sulphuric acid, washing free from acid, redissolving in water, and ammonia water, precipitating again with sulphuric acid, washing, redissolving in water and ammonia water, spreading on glass plates and scaling.

Soluble in alcohol and water. It possesses the same properties as licorice. Dose: 5 to 15 grains (0.32 to 0.97 gm.).

INULA (Elecampane).

The root of Inula Helenium (nat. ord. Compositæ), growing in Europe. Contains a glucoside Inulin, upon which its properties depend.

It is tonic and gently stimulant. Dose 20 to 60 grains (1.3 to 3.9 gm.).

IRIS (Blue Flag).

The rhizome and rootlets of Iris versicolor (nat. ord. Irideæ), growing in all parts of the United States.

Contains starch, gum, tannin, sugar, an acrid resin, fixed oil, and indications of an alkaloid.

Official in the extract and fluid extract. It is cathartic, emetic and diuretic. Dose: 10 to 20 grains (0.65 to 1.3 gm.).

EXTRACTUM IRIDIS (Extract of Iris).

Prepared by exhausting Iris No. 60 with alcohol and evaporating to a pilular consistence. Dose: 1 to 2 grains (.065 to .13 gm.).

EXTRACTUM IRIDIS FLUIDUM (Fluid Extract of Iris).

Prepared in the usual manner from Iris No. 60, using alcohol as the menstruum. Dose: 5 to 10 minims (0.3 to 0.6 C.c.).

JUGLANS (Butternut).

The bark of the root of Juglans cinerea (nat. ord. Juglandaceæ), collected in autumn; growing in Canada and the United States.

Contains Juglandic acid, resembling chrysophanic acid, upon which its properties depend. Official in the extract. It is a mild cathartic. Dose: 20 to 30 grains (1.3 to 1.95 gm.).

EXTRACTUM JUGLANDIS (Extract of Juglans).

Prepared by exhausting Juglans No. 30 with dilute alcohol and evaporating to a pilular consistence.

Dose: 5 to 10 grains (0.33 to 0.65 gm.).

LAPPA (Burdock).

The root of Arctium Lappa, and of some other species of Arctium (nat. ord. Compositæ), growing in the United States and Europe.

Contains Inulin $C_6H_{10}O_5$, a glucoside, arabin, pectin and extractive.

Official in the fluid extract. It is a diuretic and diaphoretic alterative.

EXTRACTUM LAPPÆ FLUIDUM (Fluid Extract of Lappa).

Prepared in the usual manner from burdock No. 60, using dilute alcohol as the menstruum.

Dose: 30 to 60 minims (1.37 to 3.75 C.c.).

LEPTANDRA (Culver's Root) Black Root, Culver's Physic.

The rhizome and roots of Veronica Virginica (nat. ord. Scrophularineæ), growing throughout the United States.

Contains a glucoside Leptandrin, volatile oil, extractive, tannin, gum and resin.

Official in the extract and fluid extract. Dose: 20 to 60 grains (1.3 to 3.9 gm.).

EXTRACTUM LEPTANDRÆ (Extract of Leptandra).

Prepared by exhausting leptandra No. 40, with a mixture of 750 parts alcohol and 250 parts water and evaporating to a pilular consistence.

Dose: 5 to 10 grains (0.33 to 0.65 grains).

EXTRACTUM LEPTANDRÆ FLUIDUM (Fluid Extract of Leptandra).

Prepared in the usual manner from Leptandra No. 60, using a mixture of 750 parts alcohol and 250 parts water as the menstruum. Dose: 20 to 60 minims (1.25 to 3.75 C.c.).

LACTUCARIUM (Lactucarium).

The concrete milk juice of Lactuca virosa (nat. ord. Compositæ), growing in Europe.

Contains lactusin, upon which its activity depends, resinous matter, etc.

Official in the tincture. Used as an expectorant. Dose: 10 to 20 grains (0.65 to 1.95 gm.).

TINCTURA LACTUCARII (Tincture of Lactucarium).

Prepared by macerating lactucarium with benzine to remove the resinous matter, drying, proceeding the same as in making a fluid extract, using a mixture of alcohol, water and glycerin as the menstruum.

Contains 50% lactucarium. Official in the syrup.

SYRUPUS LACTUCARII (Syrup of Lactucarium).

Prepared by mixing tincture of lactucarium with precipitated calcium phosphate, and sugar, adding water, filtering and dissolving the remainder of the sugar in the filtrate.

Used for coughs. Dose: 2 to 3 fluid drachms (7.5 to 11.25 C.c.).

MATICO (Matico).

The leaves of Piper angustifolium (nat. ord. Piperaceæ), growing in Peru and other parts of South America.

Contains a peculiar bitter principle Maticin, chlorophyll, a sort of dark green resin, brown and yellow coloring matters, gum, salts, lignin and volatile oil.

Official in the fluid extract and tincture. It is tonic and stimulant. Dose : ½ to 2 drachms (1.95 to 7.8 gm.).

EXTRACTUM MATICO FLUIDUM (Fluid Extract of Matico).

Prepared in the usual manner from matico No. 40, using a mixture of 750 parts alcohol and 250 parts water as a menstruum.

Dose : ½ to 1 fluid drachm (1.9 to 3.75 C.c.).

TINCTURA MATICO (Tincture of Matico).

Prepared by macerating and percolating matico No. 40, with dilute alcohol.

Dose : 1 fluid drachm (3.75 C.c.).

MARRUBIUM (Horehound).

The leaves and tops of Marrubium vulgare (nat. ord. Labiatæ), growing in the United States and Europe.

Contains a bitter principle Marrubiin, volatile oil, resin, tannin, lignin.

It is tonic, and in large doses laxative. Dose : 30 to 60 grains (1.95 to 3.9 gm.).

MEZEREUM (Mezereum).

The bark of Daphne Mezereum and other species of Daphne (nat. ord. Thymelaceæ), growing in Europe.

Contains a peculiar principle Daphnin, wax, an acrid resin, volatile oil, a yellow coloring matter, a gummy matter containing nitrogen, ligneous fibre and malic acid. The virtues of the recent plant reside in the Volatile Oil, which by time and exposure is changed to a resin, without losing its activity.

Official in the fluid extract, compound decoction of sarsaparilla, compound fluid extract of sarsaparilla. Dose : 10 grains (0.65 gm.).

EXTRACTUM MEZEREI FLUIDUM (Fluid Extract of Mezereum).

Prepared in the usual manner from mezereum No. 30, using alcohol as the menstruum.

It is too acrid for internal use and is only used in the form of an ointment.

PHYTOLACCÆ FRUCTUS (Poke Berry).

The fruit of Phytolacca decandra (nat. ord. Phytolaccaceæ), growing in the United States.

PHYTOLACCÆ RADIX (Poke Root).

The root of Phytolacca decandra (nat. ord. Phytolaccaceæ).

Contains a crystalline principle called Phytolaccin, tannic acid, starch, gum, resin, fixed oil, etc. The root is the most active.

Poke is emetic, purgative and somewhat narcotic. Official in the fluid extract.

EXTRACTUM PHYTOLACCÆ RADICIS (Fluid Extract of Poke Root).

Prepared in the usual manner from poke root No. 60, using a mixture of 2 parts alcohol and 1 part water as the menstruum. Dose as an alterative: 1 to 5 minims (0.06 to 0.3 C.c.).

PICROTOXINUM (Picrotoxin), $C_{30}H_{34}O_{13}$.

A neutral principle obtained from the seed of Anamirta paniculata (nat. ord. Menispermaceæ).

Sparingly soluble in water, but soluble in alcohol, in solutions of alkalies and in acids.

It is a very powerful substance, little used.

Dose: $\frac{1}{100}$ to $\frac{1}{30}$ of a grain (0.0006 to 0.0022 gm.).

PIPERINUM (Piperin) $C_{17}H_{19}NO_3$.

A neutral principle obtained from pepper and other plants of the Piperaceæ.

Insoluble in water, but soluble in alcohol.

It is a warm carminative stimulant. Dose: 1 to 6 grains (0.065 to 0.4 gm.).

PRUNUS VIRGINIANA (Wild Cherry).

The bark of Prunus serotina (nat. ord. Rosaceæ), collected in autumn, growing in the United States.

Contains a glucoside Amygdalin, another principle probably analagous to emulsin, starch, resin, tannin, gallic acid, fatty matter, lignin, red coloring matter, etc. On exposure to moisture the emulsin acts on the amygdalin, producing volatile oil and hydrocyanic acid, but these do not pre-exist in the plant ready found.

Official in the fluid extract, infusion and syrup. It is tonic, and in large doses sedative.

EXTRACTUM PRUNI VIRGINIANÆ FLUIDUM (Fluid Extract of Wild Cherry).

Prepared in the usual manner from wild cherry No. 20, using a mixture of alcohol, water and glycerin as the menstruum.

The glycerin is used to prevent precipitation. Dose: 30 to 60 minims (1.9 to 3.75 C.c.).

INFUSUM PRUNI VIRGINIANÆ (Infusion of Wild Cherry).

Prepared by macerating and percolating wild cherry with cold water. The presence of hot water would decompose the hydrocyanic acid. It is unstable and should be prepared fresh when wanted for use.

Dose: 2 to 3 fluid ounces (60 to 90 C.c.).

SYRUPUS PRUNI VIRGINIANÆ (Syrup of Wild Cherry).

Prepared by macerating and percolating wild cherry No. 20, with a mixture of glycerin and water, and dissolving sugar in the filtrate without heat.

Dose: $\frac{1}{2}$ fluid ounces (15 C.c.).

LECTURE No. 20.

PULSATILLA (Pulsatilla).

The herb of Anemone Pulsatilla and of Anemone pratensis (nat. ord. Ranunculaceæ) growing in Europe and England.

Contains a peculiar crystalline principle Anemonin.

It is rarely used.

QUASSIA (Quassia).

The wood of Picræna excelsa (nat. ord. Simarubeæ) growing in Jamaica and the Caribbean Islands.

Contains a bitter principle Quassin, upon which its action depends.

Official in the extract, fluid extract and tincture. It is tonic.

EXTRACTUM QUASSIÆ (Extract of Quassia).

Prepared by macerating and percolating quassia No. 20, with water and evaporating to a pilular consistence.

Dose: 1 to 2 grains (0.065 to 0.13 gm.).

EXTRACTUM QUASSIÆ FLUIDUM (Fluid Extract of Quassia).

Prepared in the usual manner from quassia No. 60, using a mixture of 3 parts alcohol and 6 parts water as the menstruum.

Dose: 5 to 10 minims (0.3 to 0.6 C.c.).

TINCTURA QUASSIÆ (Tincture of Quassia).

Prepared by macerating and percolating quassia No. 40, with a mixture of 350 parts alcohol and 650 parts water.

Dose: 1 fluid drachm (3.75 C.c.).

QUILLAJA (Soap Bark).

The inner bark of Quillaja Saponaria (nat. ord. Rosaceæ), growing in Peru and Chili.

Contains a glucoside Saponin.

Official in the tincture. Used in pulmonary affections, and sometimes as a substitute for senega.

TINCTURA QUILLAJÆ (Tincture of Quillaja).

Prepared by boiling coarse ground soap bark with water, adding alcohol and filtering. Used principally as an emulsifying agent.

RHEUM (Rhubarb).

The root of Rheum officinale (nat. ord. Polygonaceæ), growing in China, Turkey, England and France.

Contains a glucoside Chrysophan (which, according to Kubli, is decomposed into chrysophanic acid by the action of a ferment when the root is digested in water), extractive, tannic and gallic acids, sugar, pectin, lignin, etc. Powdered rhubarb is often prepared from inferior roots and rotten and worm-eaten roots, colored with tumeric.

Official in the extract, fluid extract, pill, compound pill, compound powder, tincture, aromatic tincture and sweet tincture.

It is carthartic and astringent. Dose: 5 to 30 grains (0.35 to 1.95 gm.).

EXTRACTUM RHEI (Extract of Rhubarb).

Prepared by macerating and percolating rhubarb No. 30, to exhaustion with a mixture of 800 parts alcohol and 200 parts water, and evaporating to a pilular consistence.

Dose: 5 to 10 grains (0.3 to 0.65 gm.).

EXTRACTUM RHEI FLUIDUM (Fluid Extract of Rhubarb).

Prepared in the usual manner from rhubarb No. 30, using a mixture of 800 parts alcohol and 200 parts water as the menstruum. Official in mixture of rhubarb and soda and the syrup.

Dose: 20 to 30 minims (0.25 to 1.9 C.c.).

MISTURA RHEI ET SODÆ (Mixture of Rhubarb and Soda).

Contains sodium bicarbonate, fluid extract of rhubarb, fluid extract of ipecac, glycerin, spirit of peppermint and water.

Dose : ½ to 1 teaspoonful (1.9 to 3.75 C.c.).

SYRUPUS RHEI (Syrup of Rhubarb).

Contains fluid extract of rhubarb, spirit of cinnamon, potassium carbonate, glycerin and syrup. Dose : 1 fluid drachm (3.7 C.c.).

PILULÆ RHEI (Pills of Rhubarb).

Contain rhubarb No. 60, soap and water. Each pill contains 3 grains of rhubarb (0.2 gm.).

PILULÆ RHEI COMPOSITA (Compound Pills of Rhubarb).

Contain rhubarb No. 60, purified aloes, myrrh, oil of peppermint and water.

Dose: 2 to 4 pills or 10 to 20 grains of the mass (0.65 to 1.3 gm.).

PULVIS RHEI COMPOSITUS (Compound Powder of Rhubarb). Gregory's Powder.

Contains rhubarb No. 60, magnesia, and ginger No. 60.

Dose : ½ to 1 drachm (1.95 to 3.9 gm.).

TINCTURA RHEI (Tincture of Rhubarb).

Prepared by macerating and percolating a mixture of rhubarb No. 40, and cardamom No. 40, with a mixture of alcohol, water and glycerin.

Dose: 1 to 2 fluid drachms (3.75 to 7.5 C.c.).

TINCTURA RHEI AROMATICA (Aromatic Tincture of Rhubarb).

Prepared by macerating and percolating a mixture of rhubarb No. 40, cassia cinnamon No. 40, clove No. 40, nutmeg No. 40, with a mixture of alcohol, water and glycerin.

Official in the aromatic syrup. Dose: 1 fluid drachm (3.75 C.c.).

SYRUPUS RHEI AROMATICUS (Aromatic Syrup of Rhubarb).

Contains aromatic tincture of rhubarb and syrup.

Dose : 1 fluid drachm (3.75 C.c.).

TINCTURA RHEI DULCIS (Sweet Tincture of Rhubarb).

Prepared by macerating and percolating a mixture of rhubarb No. 40, licorice No. 40, anise No. 40, cardamom No. 40, with a mixture of glycerin, alcohol and water.

Dose: 2 to 3 fluid drachms (7.5 to 11.25 C.c.).

RHUS TOXICODENDRON (Poison Ivy), Poison Oak.

The fresh leaves of Rhus radicans (nat. ord. Anacardieæ) growing in the United States, from Canada to Georgia.

Contains Toxicodendric acid, tannic acid, chlorophyll, wax, fixed oil, resin, sugar, albumen, pectin, starch, etc.

It is a powerful local irritant.

RUMEX (Yellow Dock).

The root of Rumex crispus and of some other species of Rumex (nat. ord. Polygonaceæ), growing in the United States and Europe.

Contains a peculiar principle Rumicin, tannin, starch, mucilage, albumen, etc.

Official in the fluid extract. It is astringent and tonic.

EXTRACTUM RUMICIS FLUIDUM (Fluid Extract of Rumex).

Prepared in the usual manner from rumex No. 40, using dilute alcohol as the menstruum.

Dose: 1 fluid drachm (3.7 C.c.).

SANTALUM RUBRUM (Red Saunders), Red Sandal Wood.

The wood of Pterocarpus santalinus (nat. ord. Leguminoseæ) growing in India and Ceylon.

Contains a coloring principle Santalin. Used principally as a coloring agent.

SALICINUM (Salicin) $C_{13}H_{15}O_7$. A neutral principal obtained from several species of Salix and Populus (nat. ord. Salicaceæ). Permanent in the air. Soluble in water and alcohol. It is a glucoside. Almost insoluble in ether and chloroform.

It is tonic. Dose: 10 to 30 grains (0.65 to 1.95 gm.).

SANTONICA (Levant Wormseed), European Wormseed.

The unexpanded flower heads of Artemisia pauciflora (nat. ord. Compositæ), growing in the Levant.

Contains Santonin, volatile oil, etc.

It is rarely used. Dose: 10 to 30 grains (0.65 to 1.95 gm.).

SANTONINUM (Santonin) $C_{15}H_{18}O_3$.

A neutral principle obtained from Santonica, and should be kept in dark amber bottles, away from the light. It is not altered by exposure to the air, but turns yellowish on exposure to light. Soluble in alcohol, chloroform and solutions of caustic alkalies.

Official in the troche. It is a vermicide. Dose: 2 to 4 grains (0.13 to 0.26 gm.).

TROCHISCI SANTONINI (Troches of Santonin). Contains santonin, sugar, tragacanth and stronger orange flower water. Each troche contains about ½ grain (0.033 gm.) of santonin. Dose: 1 to 6 troches.

SARSAPARILLA (Sarsaparilla), Jamaica Sarsaparilla.

The root of Smilax officinalis, Smilax medica, Smilax papyracea, and other undetermined species of Smilax, (nat. ord. Liliaceæ), growing in Honduras, Mexico and Jamaica.

Contains Parillin, upon which its properties depend. The quality of sarsaparilla, as found in commerce varies very much. If, on being chewed for a short time, it leaves a decidedly acrid impression in the mouth it may be considered efficient. If otherwise, it is probably inert.

Official in the fluid extract, compound decoction, and compound fluid extract.

EXTRACTUM SARSAPARILLÆ FLUIDUM (Fluid Extract of Sarsaparilla).

Prepared in the usual manner from sarsaparilla No. 30, using a mixture of 3 parts alcohol and 6 parts water as the menstruum.

Official in the compound syrup. Dose: 30 to 60 minims (1.9 to 3.75 C.c.).

SYRUPUS SARSAPARILLÆ COMPOSITUS (Compound Syrup of Sarsaparilla).

Contains a fluid extract of sarsaparilla, fluid extract of licorice, fluid extract of senna, oil of sassafras, oil of wintergreen, oil of anise, water and sugar.

Corrosive sublimate being often prescribed with this syrup is said to be reduced to calomel.

Dose: ½ fluid ounce (15 C.c.).

DECOCTUM SARSAPARILLÆ COMPOSITUM (Compound Decoctum of Sarsaparilla).

Prepared by boiling sarsaparilla (cut and bruised), guaiacum wood (rasped), with water for half an hour, adding sassafras No. 20, licorice (bruised), mezereum (cut and bruised), macerating for two hours, straining and making up to the required quantity with water.

Dose: 4 to 6 fluid ounces (120 to 180 C.c.).

EXTRACTUM SARSAPARILLÆ FLUIDUM COMPOSITUM (Compound Fluid Extract of Sarsaparilla).

Prepared in the usual manner, from a mixture of sarsaparilla No. 30, licorice No. 30, sassafras No. 30, mezereum No. 30, with a mixture of glycerin, alcohol and water.

Dose: 30 to 60 minims (1.9 to 3.75 C.c.),

SCILLA (Squill).

The bulb of Urginea maritima (nat. ord. Liliaceæ), deprived of its dry membranaceous outer scales, cut into thin slices, the central portions being rejected. It grows in Spain, France, Italy and Greece. There are two varieties in the market, the red and white. Contains a bitter principle, Scilitin. Merck obtained three compounds, namely: Scillipikrin, scillitoxin and scillin. Jamersted obtained a glucoside which he called scillain. Schmiedeberg found a peculiar carbohydrate which he called sinistrin.

Official in the vinegar, fluid extract and tincture. It is expectorant, diuretic, and in large doses emetic and purgative. Dose: 1 to 2 grains (.065 to 0.13 gm.).

ACETUM SCILLÆ (Vinegar of Squill).

Prepared by macerating squill No. 30 in dilute acetic acid for seven days, straining and filtering.

Official in the syrup. Dose: 15 to 60 minims (0.92 to 3.69 C.c).

SYRUPUS SCILLÆ (Syrup of Squill).

Prepared by heating vinegar of squill to the boiling point to coagulate the albuminous matter, filtering, dissolving sugar in the filtrate, and adding water to make the required quantity.

Incompatible with ammonium carbonate. Dose: 1 fluid drachm (3.7 C.c.).

EXTRACTUM SCILLÆ FLUIDUM (Fluid Extract of Squill).

Prepared in the usual manner from squill No. 20, with a mixture of 750 parts alcohol and 250 parts water as the menstruum.

Official in compound syrup of squill.

SYRUPUS SCILLÆ COMPOSITUS (Compound Syrup of Squill).

Prepared by mixing fluid extract of squill and fluid extract of senega in a tared capsule, evaporating, mixing the residue with water, adding precipitated calcium phosphate, filtering, adding tartar emetic dissolved in water, dissolving sugar in the solution by agitation, without heat, and adding water to make the required quantity.

Each fluid ounce contains 1 grain of tartar emetic. Dose: 20 to 30 drops (1.25 to 1.9 C.c.). For children: 10 to 60 drops (0.6 to 3.7 C.c.).

TINCTURA SCILLÆ (Tincture of Squill).

Prepared by macerating and percolating squill No. 30, with a mixture of 750 parts alcohol and 250 parts water.

Dose: 10 to 20 minims (0.6 to 1.25 C.c.).

SCUTELLARIA (Scullcap).

The herb of Scutellaria lateriflora (nat. ord. Labiatæ), growing in the United States.

It is probably of no remedial effect. Scutellarin which it is said to contain is not a pure proximate principle, but is obtained by mixing a concentrated tincture with water, precipitating with alum, washing and drying, and varies in dose from 1 to 3 or 4 grains (0.065 to 0.20 or 0.26 gm.). Official in the fluid extract.

EXTRACTUM SCUTELLARIÆ FLUIDUM (Fluid Extract of Scutellaria).

Prepared in the usual manner from scutellaria No. 40, using dilute alcohol as the menstruum.

Dose: $\frac{1}{2}$ to 1 fluid drachm (1.9 to 3.75 C.c.).

SENEGA (Senega), Senega Snakeroot.

The root of Polygala Senega (nat. ord. Polygaleæ), growing in the United States.

Contains a glucoside Senegin or Polygalic Acid, now recognized as Saponin, fixed oil, resin, traces of volatile oil (a mixture of valerianic ether and methyl-salicylate), sugar, yellow coloring matter and malates.

Official in the fluid extract. It is a stimulant, expectorant and diuretic, and in large doses, emetic and cathartic. Dose: 10 to 20 grains (0.65 to 1.3 gm.).

EXTRACTUM SENEGÆ FLUIDUM (Fluid Extract of Senega).

Prepared in the usual manner from senega No. 40, using a mixture of alcohol, water and ammonia water as the menstruum.

On account of the pectionus bodies in senega it is liable to gelatinize, and the ammonia water prevents this.

Official in the syrup and compound syrup of squill.

SYRUPUS SENEGÆ (Syrup of Senega).

Contains fluid extract of senega, ammonia water, sugar and water.

Dose: 1 or 2 fluid drachms (3.7 to 7.5 C.c.).

SENNA (Senna).

The leaflets of Cassia acutifolia and Cassia angustifolia (nat. ord. Leguminosæ), growing in Egypt and Africa. There are four varieties in the market: Alexandria, Tripoli, India and Mecca. Only the Alexandria and India are recognized by the U. S. P.

Contains a glucoside Cathartic acid, chrysophanic acid, sennacrol, sennapicrin and a saccharine principle.

Official in the confection, fluid extract, compound infusion, syrup and compound licorice powder. It is a purgative. Dose: ½ to 2 drachms (1.95 to 7.8 gm.).

CONFECTIO SENNÆ (Confection of Senna).

Contains senna No. 60, cassia fistula (bruised), tamarind, prune, fig, sugar, oil of coriander and water. It is a pleasant laxative. Dose: 2 drachms (7.8 gm.).

EXTRACTUM SENNÆ FLUIDUM (Fluid Extract of Senna).

Prepared in the usual manner from senna No. 30, using dilute alcohol as the menstruum.

Official in compound syrup of sarsaparilla. Dose: 1 to 4 fluid drachms (3.75 to 15 C.c.).

INFUSUM SENNÆ COMPOSITUM (Compound Infusion of Senna), Black Draught.

Prepared by pouring boiling water upon a mixture of senna and fennel, macerating, straining, expressing, dissolving magnesium sulphate and manna in the expressed liquid, and adding water to make the sufficient quantity. Dose: 4 fluid ounces (118 C.c.).

SYRUPUS SENNÆ (Syrup of Senna).

Prepared by digesting senna in boiling water, expressing, straining, mixing with alcohol in which oil or coriander has been dissolved, decanting off the clear liquid and dissolving sugar in it without the aid of heat.

Dose: 2 to 4 fluid drachms (7.5 to 15 C.c.).

SINAPIS ALBA (White Mustard).

The seed of Brassica alba (nat. ord. Cruciferæ).

Contains a glucoside Sinalbin and a ferment Myrosin. In the presence of water the ferment acts on the glucoside, forming Acrinyl Sulphocyanate, a body forming part of the essential oil of mustard paste.

It is a laxative, stimulant and rubifacient.

SINAPIS NIGRA (Black Mustard).

The seed of Brassica nigra (nat. ord. Cruciferæ), growing in all parts of the United States and Europe.

Contains a glucoside Sinigrin (potassium myronate), and a ferment Myrosin, which in the presence of water react, forming Allyl Sulphocyanate, which is the chief part of the pungent oil of mustard paste.

Both the white and black mustard are used in the powdered state and are often adulterated with wheat flour, colored with tumeric and red pepper is often added. They both yield a fixed oil. Used for the same purpose as the white. Official in the paper.

CHARTA SINAPIS (Mustard Paper).

Prepared by percolating black mustard No. 60 with benzin and drying to remove the fixed oil, mixing with India Rubber which has been dissolved in a mixture of carbon disulphide and benzin, and spreading on paper or cloth. It is a rubifacient.

STROPHANTHUS (Strophanthus).

The seed of Strophanthus hispadus (nat ord. Apocynaceæ), growing in Africa and Asia, deprived of its long awn.

Contains a glucoside Strophanthin and an alkaloid inæine. Official in the tincture. It is a cardiac stimulant.

TINCTURA STOPHANTHUS (Tincture of Strophanthus).

Prepared by digesting strophanthus No. 30 with a mixture of 650 parts alcohol and 350 parts water and percolating.

Dose: 5 to 15 minims (0.3 to 0.9 C.c.).

TARAXACUM (Dandelion).

The root of Taraxacum officinale (nat. ord. Compositæ), growing all over the world.

Contains a bitter principle Taraxacin, inulin, sugar, etc. Official in the extract, and fluid extract. It is tonic, diuretic and aperient.

EXTRACTUM TARAXACI (Extract of Taraxacum).

Prepared by bruising fresh dandelion with water until it is reduced to pulp, expressing and evaporating to a syrupy consistence; it should be kept in close vessels and its surface covered with a cloth, which should be moistened occasionally with a little ether or chloroform. The expressed juice yields from 11 to 25% of extract.

Dose: 20 to 60 grains (1.3 to 3.95 gm.).

EXTRACTUM TARAXACI FLUIDUM (Fluid Extract of Taraxacum).

Prepared in the usual manner from dandelion No. 30, using dilute alcohol as the menstruum.

Dose: 1 to 3 fluid drachms (3.75 to 11.25 C.c.).

TRITICUM (Couch Grass), Quitch, Dog-grass, Quick-grass.

The rhizome of Agropyrum repens (nat. ord Gramineæ), gathered in the spring and deprived of the roots. It grows in the United States.

Official in the fluid extract. Used in bladder and kidney troubles.

EXTRACTUM TRITICI FLUIDUM (Fluid Extract of Triticum).

Prepared by percolating triticum to exhaustion with boiling water, evaporaring to a small bulk, adding alcohol, filtering and making up to the required quantity with a mixture of 1 part alcohol and 3 parts water.

Dose: 3 to 6 fluid drachms (11.25 to 22.5 C.c.).

UVA URSI (Bearberry).

The leaves of Arctostaphylos Uva Ursi (nat. ord. Ericaceæ), growing in the United States, Europe and Asia.

Contains a glucoside Arbutin, tannic and gallic acids, resin, gum, fatty matter, etc. Official in the extract and fluid extract. It is astringent, tonic and diuretic. Dose: 20 to 60 grains (1.3 to 3.9 gm.).

EXTRACTUM UVÆ URSI (Extract of Uva Ursi).

Prepared by exhausting uva ursi No. 30 with a mixture of 2 parts alcohol and 5 parts water, and evaporating to pilular consistence.

Dose: ½ to 1 drachm (1.9 to 3.75 gm.).

EXTRACTUM UVÆ URSI FLUIDUM (Fluid Extract of Uva Ursi).

Prepared in the usual manner from uva ursi No. 30, using a mixture of alcohol, glycerin and water as the menstruum.

Dose: 30 to 60 minims (1.9 to 3.75 C.c.).

VANILLA (Vanilla).

The fruit of Vanilla planifolia (nat. ord. Orchideæ), growing in Mexico and South America.

Contains Vanillin $C_8H_5O_3$, fixed oil, resin, wax, tannic acid, sugar, starch and benzoic acid. There are five varieties in the market, namely: Mexican (first quality), Bourbon, Mauritius and Leychelles (inferior Bourbon of the trade), South American and Tahiti (transplanted Mexican). Vanillin may be prepared synthetically from coniferin, eugenol or guaiacol.

Official in the tincture. Used principally as a flavoring agent.

LECTURE NO. 21.

TINCTURA VANILLÆ (Tincture of Vanilla).

Prepared by macerating vanilla in a mixture of 650 parts alcohol and 350 parts water for twelve hours, draining, beating with sugar, pouring over it the drained liquid and percolating with the menstruum to the required quantity. Used as a flavoring agent.

VIBURNUM PRUNIFOLIUM (Black Haw), Sloe, Stagbush.

The bark of Viburnum prunifolium (nat. ord. Caprifoliaceæ), growing in the United States.

Contains a neutral principle Viburnin, a brownish resinous body, valerianic acid, tannic acid, oxalic acid, citric acid, malic acid, sulphates, etc. Official in the fluid extract. It is nervine, antispasmodic, astringent, diuretic and tonic.

EXTRACTUM VIBURNI PRUNIFOLII FLUIDUM (Fluid Extract of Black Haw).

Prepared in the usual manner from black haw No. 60, using a mixture of 750 parts of alcohol and 250 parts water as the menstruum.

Dose: $\frac{1}{2}$ to 1 fluid drachm (1.9 to 3.75 C.c.).

XANTHOXYLUM (Prickly Ash), Toothache Tree, Angelica Tree, Suterberry, Pepper Wood, Tea Ash.

The bark of Xanthoxylum Americanum and of Xanthoxylum Clava-Herculis (nat. ord. Rutaceæ), growing in the United States.

Contains a crystalline principle Xanthoxylin, fixed oil, berberine, etc. Official in the fluid extract. It is stimulant.

EXTRACTUM XANTHOXYLI FLUIDUM (Fluid Extract of Xanthoxylum).

Prepared in the usual manner from xanthoxylum No. 40, using alcohol as the menstruum. Dose: $\frac{1}{2}$ to 1 drachm (1.9 to 3.75 C.c.).

ZEA (Corn Silk).

The styles and stigmas of Zea Mays (nat. ord. Gramineæ), growing in the United States.

Contains Maizenic Acid, fixed oil, resin, chlorophyll, sugar, gum, extractive, albuminoids, phlobaphene salt, cellulose and water.

It is a mild stimulant diuretic.

ALKALOIDS.

ACONITUM (Aconite), Monkshood, Wolfsbane.

The tuber of Aconitum Napellus (nat. ord. Ranunculaceæ), growing in Germany, France and Switzerland, which should be gathered in autumn or winter, after the leaves have fallen, and is not perfect until the second year. The root is much more active than the leaves.

Contains an alkaloid Aconitia, $C_{33}H_{45}NO_{12}$ (Dunstan) aconine $C_{26}H_{41}NO_{11}$, Napelline and homo-napelline, aconitic acid, etc.

Aconite was official in the 1880 U. S. P., but on account of its not being a definite proximate principle as obtained by the process of the 1870 U. S. P., and the variation in strength making it dangerous to administer, it was dropped. The antidotes to aconite poisoning are whiskey and brandy internally, tincture of digitalis and strychnine hypodermically, given separately. It is a cardiac sedative. Official in the extract, fluid extract and tincture. Dose : 1 to 2 grains (.065 to 0.13 gm.).

EXTRACTUM ACONITI (Extract of Aconite).

Prepared by macerating aconite No. 60 with alcohol, percolating to exhaustion and evaporating to a pilular consistence.

Dose : $\frac{1}{8}$ to $\frac{1}{4}$ of a grain (0.01 to 0.016 gm.).

EXTRACTUM ACONITI FLUIDUM (Fluid Extract of Aconite).

Prepared in the usual manner from aconite No. 60, using a mixture of 3 parts alcohol and 1 part water as the menstruum.

Dose : $\frac{1}{2}$ to 1 minims (0.03 to 0.06 C.c.).

TINCTURA ACONITI (Tincture of Aconite).

Prepared by macerating and percolating aconite No. 60 with a mixture of 7 parts alcohol and 3 parts water as the menstruum.

Dose : 1 to 3 drops (0.06 to 0.18 C.c.).

ARNICÆ FLORES (Arnica Flowers), Leopard's Bane.

The flower heads of Arnica montana (nat. ord. Compositæ), growing in the United States, Europe and Siberia.

Contains an alkaloid Arnicine $C_{35}H_{54}O_7$. Official in the tincture. Used as a sternutatory.

TINCTURA ARNICÆ (Tincture of Arnica).

Prepared by macerating and percolating arnica flowers with diluted alcohol.

Used as an application to sprains and bruises.

ARNICÆ RADIX (Arnica Root).

The rhizome and roots of Arnica montana (nat. ord. Compositæ).

Contains an alkaloid Arnicine, volatile oil, inulin, etc. Official in the extract, fluid extract, and tincture.

EXTRACTUM ARNICÆ RADICIS (Extract of Arnica Root).

Prepared by macerating and percolating arnica root No. 60 with dilute alcohol and evaporating to a pilular consistence.

Official in plaster. Dose : 3 to 5 grains (0.2 to 0.33 gm.).

EMPLASTRUM ARNICÆ (Arnica Plaster).

Contains extract of arnica root and resin plaster.

Used for sprains and bruises.

EXTRACTUM ARNICÆ RADICIS FLUIDUM (Fluid Extract of Arnica Root).

Prepared in the usual manner from arnica root No. 60, using a mixture of 750 parts alcohol and 250 parts water, as a menstruum.

Dose : 5 to 10 minims (0.3 to 0.6 C.c.).

TINCTURA ARNICÆ RADICIS (Tincture of Arnica Root).

Prepared by macerating and percolating arnica root No. 40 with a mixture of 650 parts alcohol and 350 parts water.

Dose : 20 to 30 minims (1.25 to 1.9 C.c.).

ASPIDOSPERMA (Quebracho).

The bark of Aspidosperma Quebracho-blanco (nat. ord. Apocynaceæ), growing in South America.

Contains an alkaloid Aspidospermine $C_{22}H_{30}N_2O_2$, and four other alkaloids : Aspidospermatine, aspidosamine, quebrachine, and quebrachamine. Official in the fluid extract. It is anti-periodic.

EXTRACTUM ASPIDOSPERMATIS FLUIDUM (Fluid Extract of Aspidosperma).

Prepared in the usual manner from aspidosperma No. 60, using a mixture of alcohol, water and glycerin as the menstruum.

Dose : 15 to 60 minims (0.9 to 3.75 C.c.).

APOMORPHINÆ HYDROCHLORAS (Apormorphine Hydroclorate), $C_{17}H_{17}NO_2HCl$.

The hydrochlorate of an artificial alkaloid prepared from morphine or codeine by heating in sealed tubes with hydrochloric acid and should be kept in small amber-colored vials. It acquires a greenish tint on exposure to light and air. Soluble in alcohol and water, but little soluble in ether or chloroform.

On adding sodium bicarbonate solution to an aqueous solution of the salt, it-throws down the white amorphous alkaloid, which soon turns green on exposure to the air, and imparts a violet or blue color to chloroform, in which it is very soluble (difference from morphine).

It is emetic. Hypodermic dose : $\frac{1}{15}$ to $\frac{1}{10}$ of a gr. (0.004 to 0 006 gm.).

BELLADONNÆ FOLIA (Belladonna Leaves), Deadly Nightshade.

The leaves of Atropa Belladonna (nat. ord. Solanaceæ), growing in Europe and cultivated in this country.

Contains an alkaloid Atropine $C_{17}H_{23}NO_3$ and another one hyoscyamine. Official in the alcoholic extract and tincture. Its action is the same as that of atropine.

EXTRACTUM BELLADONNÆ FOLIORUM ALCOHOLICUM (Alcoholic Extract of Belladonna Leaves).

Prepared by macerating and percolating belladonna leaves No. 60 with a mixture of 2 parts alcohol and 1 part water until exhausted, and evaporating to a pilular consistence.

Official in the plaster and ointment. Dose : $\frac{1}{6}$ to $\frac{1}{3}$ of a grain (0.01 to 0.021 gm.).

EMPLASTRUM BELLADONNÆ (Belladonna Plaster)..

Contains alcoholic extract of belladonna, resin plaster and soap plaster.

Used in rheumatic and neuralgic pains.

UNGUENTUM BELLADONNÆ (Belladonna Ointment)

Contains alcholic extract of belladonna rubbed with a little dilute alcohol and mixed with benzoinated lard.

TINCTURA BELLADONNÆ FOLIORUM (Tincture of Belladonna Leaves).

Prepared by macerating and percolating belladonna leaves No. 60 with dilute alcohol. Dose: 15 to 30 drops (.9 to 1.9 C.c.).

BELLADONNÆ RADIX (Belladonna Root), Deadly Nightshade.

The root of Atropa Belladonna (nat. ord. Solanaceæ). Contains three alkaloids: Atropine $C_{17}H_{23}NO_3$, Belladonnine $C_{17}H_{23}NO_4$, and Hyoscyamine. The root is more powerful than the leaves. Official in the fluid extract. Its uses are the same as those of the leaves.

EXTRACTUM BELLADONNÆ RADICIS FLUIDUM (Fluid Extract of Belladonna Root).

Prepared in the usual manner from belladonna root No. 60, using a mixture of 8 parts alcohol and 2 parts water as the menstruum. Official in belladonna liniment. Dose: 1 to 2 minims (0.06 to 0.12 C.c.).

LINIMENTUM BELLADONNÆ (Belladonna Liniment).

Prepared by dissolving camphor in fluid extract of Belladonna.

ATROPINA (Atropine) $C_{17}H_{23}NO_3$.

An alkaloid obtained from belladonna, and as it occurs in commerce is always accompanied by a small proportion of hyoscyamine extracted along with it, from which it cannot be readily separated.

It gradually assumes a yellowish tint on exposure to the air. Sparingly soluble in water, but very soluble in alcohol, ether and chloroform. Its aqueous solution, or that of any of its salts, is not precipitated by platinic chloride test solution (difference from most other alkaloids).

It is inflammable, giving off an odor like that of benzoic acid. It is a midriatic, sedative and stimulant to the respiratory centres.

The treatment of poisoning by atropine consists in emptying the stomach, cold applications to the head, opium preparations internally and stimulants. Dose: $\frac{1}{75}$ of a grain (0.00086 gm.) by mouth, and $\frac{1}{100}$ of a grain (0.00065 gm.) hypodermically.

ATROPINÆ SULPHAS (Atropine Sulphate) $(C_{17}H_{23}NO_3)_2, H_2SO_4$.

Prepared by dissolving atropine in a mixture of sulphuric acid and water and evaporating.

Permanent in the air. Soluble in water and alcohol; sparingly soluble in ether and chloroform.

Used for the same purposes and in the same doses as atropine.

CALUMBA (Columbo) Colombo.

The root of Jeteorhiza palmata (nat. ord. Menispermaceæ), growing in Africa. Contains an alkaloid Berberine $C_{20}H_{17}NO_4$, columbin, columbic acid $C_{22}H_{24}O_7$, etc. Official in the fluid extract and tincture. It is a tonic. Dose: 10 to 30 grains (0.65 to 1.95 gm.).

EXTRACTUM CALUMBÆ FLUIDUM (Fluid Extract of Columbo).

Prepared in the usual manner from columbo No. 20, using a mixture of 75 parts alcohol and 25 parts water as the menstruum.

Dose: 15 to 30 minims (0.9 to 1.9 C.c.).

TINCTURA CALUMBÆ (Tincture of Columbo).

Prepared by macerating and percolating columbo No. 20 with a mixture of 600 parts alcohol and 400 parts water.

Dose: 1 to 4 fluid drachms (3.7 to 15 C.c.).

CHELIDONIUM (Celandine) Tetterwort.

The entire plant of Chelidonium majus (nat ord. Papaveraceæ), growing in the United States and Europe. Contains an alkaloid Chelidonine $C_{20}H_{19}NO_5$, H_2O, and four other alkaloids: Chelerythrine, a-homochelidonine, b-homochelidonine, protopine.

It is an acrid purgative.

Dose: 30 to 60 grains (1.9 to 3.9 gm.).

CINCHONA (Cinchona) Cinchona Bark, Peruvian Bark.

The bark of Cinchona Calisaya, Cinchona officinalis, and of hybrids of these and of other species of Cinchona (nat. ord. Rubiaceæ) containing not less than 5% of total alkaloids, and at least 2.5% of quinine. It grows in South America and contains 22 alkaloids of which the following are official: Quinine $C_{20}H_{24}N_2O_2$, Quinidine $C_{20}H_{24}N_2O_2$, Cinchonine $C_{19}H_{22}N_2O$, and Cinchonidine $C_{19}H_{22}N_2O$.

Official in the fluid extract, extract infusion and tincture. It is antiperiodic.

EXTRACTUM CINCHONÆ (Extract of Cinchona).

Prepared by macerating and percolating cinchona No. 60 with a mixture of 3 parts alcohol and 1 part water to exhaustion, distilling off the alcohol and evaporating to a pilular consistence.

Dose: 10 to 30 grains (0.65 to 1.9 gm.).

EXTRACTUM CINCHONÆ FLUIDUM (Fluid Extract of Cinchona).

Prepared in the usual manner from cinchona No. 60, using a mixture of glycerin, alcohol and water as the menstruum. Dose: 1 fluid drachm (3.75 C.c.).

INFUSUM CINCHONÆ (Infusion of Cinchona) Infusion of Yellow Bark.

Prepared by percolating cinchona No. 40 with a mixture of water and aromatic sulphuric acid. The acid is used to extract the alkaloids.

Incompatible with the alkalies, alkaline carbonates, alkaline earths, soluble salts of iron, zinc and silver, mercuric chloride, arsenous acid, tartar emetic, infusions and decoctions of galls, chamomile, columbo, cascarilla, horseradish, cloves, catechu, orange peel, foxglove, senna, rhubarb and valerian. Dose: 2 fluid ounces (60 C.c.).

TINCTURA CINCHONÆ (Tincture of Cinchona) Tincture of Yellow Cinchona.

Prepared by macerating and percolating cinchona No. 60 with a mixture of glycerin, alcohol and water. Dose: 1 to 4 fluid drachms (3.7 to 15 C.c.).

CINCHONA RUBRA (Red Cinchona) Red Bark.

The bark of Cinchona succirubra (nat. ord. Rubiaceæ) and should contain not less than 5% of total alkaloids.

Official in the compound tincture.

TINCTURA CINCHONÆ COMPOSITA (Compound Tincture of Cinchona).

Prepared by macerating and percolating a mixture of red cinchona No. 60, bitter orange peel No. 60, and serpentaria No. 60, with a mixture of glycerin, alcohol and water. Dose: 1 to 4 fluid drachms (3.7 to 15 C.c.).

CINCHONIDINÆ SULPHAS (Cinchonidine Sulphate), $(C_{19}H_{22}N_2O)_2 H_2SO_4, 3H_2O$.

The neutral sulphate of an alkaloid obtained from the bark of various species of cinchona, prepared from the mother liquors obtained in the manufacture of quinine sulphate, and is separated from the sulphates of the other alkaloids by fractional crystallization.

It is slightly efflorescent on exposure to the air. Soluble in water and alcohol; sparingly soluble in chloroform and ether. The presence of sulphates of other cinchona alkaloids increase its solubility in ether and chloroform.

A solution of the salt (1 in 1000) in dilute sulphuric acid should not exhibit more than a faint blue florescence (absence of more than traces of the sulphates of quinine or quinidine).

Its action is the same as that of quinine only less powerful.

Dose: 3 to 30 grains (0.19 to 1.95 gm.).

CINCHONINA (Cinchonine) $C_{19}H_{22}N_2O$.

An alkaloid obtained from the bark of various species of Cinchona. Prepared by precipitating an aqueous solution of cinchonine sulphate with ammonia water. Permanent in the air. Sparingly soluble in water, partially soluble in alcohol, ether and chloroform. It may be distinguished from quinine by failing to respond to the Thalleioquin test (see quinine). The test for quinine and quinidine is the same as in cinchonidine sulphate.

Its action is the same as that of quinine only less powerful.

Dose: 3 to 30 grains (0.19 to 1.95 gm.).

CINCHONINÆ SULPHAS (Cinchonine Sulphate) $(C_{19}H_{22}N_2O)_2 H_2SO_4, 2 H_2O$.

The sulphate of an alkaloid obtained from the bark of various species of cinchona. Prepared from the mother liquor left after extracting quinine sulphate. There are two cinchonine sulphates; the neutral and the acid sulphate, the official salt being the neutral.

Permanent in the air. Soluble in water, alcohol, and chloroform, insoluble in ether. The test for quinine and quinidine is the same as in cinchonine. The acid sulphate or bisulphate ($C_{19}H_{22}N_2O, H_2SO_4, 3 H_2O$) may be prepared by adding sulphuric acid to the neutral sulphate.

Used for the same purpose as quinine, only is less powerful.

Dose : 3 to 30 grains (0.19 to 1.95 gm.).

QUINIDINÆ SULPHAS (Quinidine Sulphate), $(C_{20}H_{24}N_2O_2)_2H_2SO_4$, $2 H_2O$.

The neutral sulphate of an alkaloid obtained from the bark of several species of cinchona, and should be kept in well-stoppered bottles, in a dark place. The alkaloid quinidine is isomeric with quinine.

Permanent in the air. Soluble in water and more so in alcohol, almost insoluble in ether.

A cold, saturated aqueous solution of the salt yields a white precipitate with potassium iodide test solution (difference from quinine sulphate). The salt does not produce a red color when mixed with nitric acid (difference from morphine).

Its action is the same as that of quinine only less powerful. Dose : 20 to 60 grains (1.3 to 3.9 gm.).

QUININA (Quinine) $C_{20}H_{24}N_2O_2, 3H_2O$.

An alkaloid obtained from the bark of several species of cinchona, and should be kept in well-stoppered bottles, in a dark place. Prepared by adding to a solution of quinine sulphate, a quantity of ammonia water or solution of soda just sufficient to precipitate the alkaloid.

Permanent in the air. Sparingly soluble in water; soluble in alcohol, ether, chloroform, benzin, benzol, ammonia water, and dilute acids. Quinine and its salts are distinguished from all other alkaloids and their salts, except quinidine and quinicine, by the beautiful emerald green color which results when their solutions are treated first with solution of chlorine and then with ammonia, and which changes to a white or violet upon saturation with a dilute acid (THALLEIOQUIN TEST). A solution of quinine in dilute sulphuric acid has a vivid blue florescence.

Quinine does not produce a red color when treated with nitric acid (difference from morphine).

It is a febrifuge, antiperiodic, antipyretic and uterine stimulant. Dose same as that of quinine sulphate. Official in citrate of iron and quinine and soluble citrate of iron and quinine.

QUININÆ BISULPHAS (Quinine Bisulphate) $C_{20}H_{24}N_2O_2H_2SO_4$, $7H_2O$.

Prepared by suspending quinine sulphate in water, adding a molecule of sulphuric acid, filtering and crystallizing, and should be kept in well-stoppered bottles, in a dark place.

It possesses great advantages over quinine sulphate in being soluble in water. Efflorescent on exposure to the air. Soluble in water and alcohol.

The dose is about 15% greater than that of the sulphate on account of the great proportion of acid in it.

QUININÆ HYDROBROMAS (Quinine Hydrobromate) $C_{20}H_{24}N_2O_2$ HBr, H_2O.

Prepared by suspending quinine sulphate in water, boiling, adding barium bromide dissolved in water, filtering, evaporating and crystallizing, and should be kept in well-stoppered bottles, in a dark place.

It is liable to lose water on exposure to warm or dry air. Soluble in water, alcohol, ether and choloroform.

Dose the same as that of quinine sulphate.

QUININÆ HYDROCHLORAS (Quinine Hydrochlorate) $C_{20}H_{24}N_2O_2HCl$, $2H_2O$.

Obtained by the same process and from the same sources as quinine sulphate, the separated alkaloid being neutralized by hydrochloric acid and should be kept in well-stoppered bottles, in a dark place.

It is liable to lose water when exposed to warm air. Soluble in water, alcohol and chloroform.

Dose the same as that of quinine sulphate.

QUININÆ SULPHAS (Quinine Sulphate) $(C_{20}H_{24}N_2O_2)_2H_2SO_4$, $7H_2O$.

Prepared by acting on an acidulated aqueous infusion of cinchona with an alkali, neutralizing the alkaloid with sulphuric acid, purifying, and should be kept in well-stoppered bottles, in a dark place.

It is liable to lose water when exposed to warm air and to absorb water when exposed to moist air, and to become colored on exposure to light. Sparingly soluble in water; soluble in alcohol, glycerin and dilute acids; sparingly soluble in chloroform.

Incompatible with the alkalies and their carbonates, the alkaline earths, potassa, soda, ammonia, astringent infusions, soluble salts of lead, soluble salts of acetic, oxalic, tartaric and gallic acids. Sometimes adulterated with calcium sulphate, other alkaline earthy salts, sulphates of other cinchona alkaloids, gum, sugar, mannite, starch, stearin, caffeine, salicin, etc.

Official in the syrup of phosphates of iron, quinine and strychnine.

Dose: 1 to 30 grains (0.065 to 1.95 gm.).

QUININÆ VALERIANAS (Quinine Valerianate) $C_{20}H_{24}N_2O_2C_5H_{10}O_2$, H_2O.

Prepared by decomposing quinine sulphate by means of ammonia, and combining directly with valerianic acid. It should be kept in well-stoppered bottles, in a dark place. Permanent in the air. Soluble in water and more so in alcohol. Dose: 1 to 2 grains (0.065 to 0.13 gm.).

LECTURE NO. 22.

COCA (Coca).

The leaves of Erythroxylon Coca (nat. ord. Lineæ) growing in South America.

Contains an alkaloid Cocaine, resin, tannin, an aromatic principle, extractive, chlorophyll, etc.

Official in the fluid extract. It is a nerve stimulant. Dose: 30 to 60 grains (1.95 to 3.9 gm.).

EXTRACTUM COCÆ FLUIDUM (Fluid Extract of Coca).

Prepared in the usual manner from coca No. 40, using dilute alcohol as the menstruum.

Used as a nerve stimulant. Dose: 20 to 60 minims (1.25 to 3.75 C.c.).

COCAINÆ HYDROCHLORAS (Cocaine Hydrochlorate) $C_{17}H_{21}NO_4$, HCl.

The hydrochlorate of an alkaloid obtained from coca.

Permanent in the air. Soluble in water, alcohol, chloroform; sparingly soluble in ether.

It is a cerebral stimulant and is often used as a local anæsthetic.

Dose: $\frac{1}{2}$ to 1 grain (0.032 to 0.065 gm.).

COLCHICI RADIX (Colchicum Root) Meadow Saffron Root.

The corm of Colchicum autumnale (nat. ord. Liliaceæ) growing in Europe and Northern Africa.

Contains an alkaloid Colchicine upon which its properties depend.

It should be collected from the early part of June to the middle of August. If collected early in the spring it is too young to have fully developed its peculiar properties, and later in the Fall it has become exhausted by nourishing the new plant.

Official in the extract, fluid extract and wine. Used principally in the treatment of gout and rheumatism.

EXTRACTUM COLCHICI RADICIS (Extract of Colchicum Root).

Prepared by macerating and percolating colchicum root No. 60 with a mixture of acetic acid and water, and evaporating to a pilular consistence. The acid is used to render the alkaloid more soluble.

Dose: 1 to 2 grains (0.065 to 0.13 gm.).

EXTRACTUM COLCHICI RADICIS FLUIDUM (Fluid Extract of Colchicum Root).

Prepared in the usual manner from colchicum root No. 60, using a mixture of 2 parts alcohol and one part water as the menstruum.

Dose: 2 to 8 minims (0.12 to 0.5 C.c.).

VINUM COLCHICI RADICIS (Wine of Colchicum Root).

Prepared by percolating colchicum root No. 30 with a mixture of white wine and alcohol, and contains 40%.

Dose. 10 to 60 minims (0.6 to 3.75 C.c.).

COLCHICI SEMEN (Colchicum Seed).

The seed of Colchicum autumnale (nat. ord. Liliaceæ) and should be collected about the end of July or beginning of August.

Contains an alkaloid Colchicine upon which its properties depend.

Official in the fluid extract, tincture and wine.

EXTRACTUM COLCHICI SEMINIS FLUIDUM (Fluid Extract Colchicum Seed).

Prepared in the usual manner from colchicum seed No. 30, using a mixture of 2 parts alcohol and 1 part water as the menstruum.

Dose: 2 to 8 minims (0.12 to 0.5 C.c.).

TINCTURA COLCHICI SEMINIS (Tincture of Colchicum Seed).

Prepared by macerating and percolating colchicum seed No. 30 with a mixture of 6 parts alcohol and 4 parts water. Dose: ½ to 2 fluid drachms (1.9 to 7.5 C.c.).

VINUM COLCHICI SEMINIS (Wine of Colchicum Seed).

Prepared by macerating colchicum seed No. 30 for seven days with a mixture of white wine and alcohol, and filtering, and contains 15%.

Dose: ½ to 2 fluid drachms (1.9 to 7.5 C.c.).

CONIUM (Hemlock).

The full-grown fruit of Conium maculatum (nat. ord. Umbelliferæ), gathered while yet green, and growing in the United States and Europe.

Contains a volatile alkaloid called Coniine $C_8H_{17}N$, and a small quantity of volatile oil.

Official in the extract and fluid extract. It is narcotic.

EXTRACTUM CONII (Extract of Conium).

Prepared by macerating and percolating conium No. 40 with a mixture of acetic acid and diluted alcohol and evaporating to a pilular consistence. The acid is used to fix the alkaloid coniine.

Dose: ½ to 1 grain (0.03 to 0.065 gm.).

EXTRACTUM CONII FLUIDUM (Fluid Extract of Conium), Fluid Extract of Conium Seed.

Prepared in the usual manner from Conium No. 40, using a mixture of acetic acid and diluted alcohol as the menstruum.

Dose: 5 minims (0.3 C.c.).

ERGOTA (Ergot), Spurred Rye.

The schlerotium of Claviceps purpurea replacing the grain of the rye, Secale cereale (nat. ord. Gramineæ) growing in all parts of the United States. It should be only moderately dried, and preserved in a close vessel with a few drops of chloroform dropped upon it, from time to time, to prevent the development of insects. It should not be used when more than one year old.

In all the Gramineæ, or grass tribe, and in some of the Cyperaceæ, the place of the seeds is sometimes occupied by a morbid growth which, from the resemblance to the spur of a cock, has received the name of ergot, adopted from the French. It is most frequent in the rye.

Contains, according to Menzel, two alkaloids Ecboline and Ergotine, which are its active principles. Tanret claims that ergot contains an unstable alkaloid Ergotinine, which is accompanied by a volatile camphoraceous substance. Dragendorff claims that it contains Schlerotic or Schlerotinic Acid, upon which its properties depend. It also contains Ergotin, a resinous substance, and a fixed oil.

Ergot, when kept perfectly dry and in well-stoppered bottles, will keep its virtues for a long time, but when exposed to air and moisture it quickly undergoes a change and deteriorates. Official in the extract, fluid extract and wine. It is an oxytoxic.

EXTRACTUM ERGOTÆ (Extract of Ergot).

Prepared by evaporating fluid extract of ergot to a pilular consistence at a temperature not exceeding 50°C (122°F).

Dose: 5 to 30 grains (0.33 to 1.9 gm.).

EXTRACTUM ERGOTÆ FLUIDUM (Fluid Extract of Ergot).

Prepared in the usual manner from ergot No. 60, using a mixture of dilute alcohol and acetic acid as the menstruum. The acid is used to fix the alkaloids. Dose: ½ fluid drachm to ½ fluid ounce (1.9 to 15 C.c.).

VINUM ERGOTÆ (Wine of Ergot).

Prepared by percolating ergot No. 30 with a mixture of white wine and alcohol.

Dose: 1 to 4 fluid drachms (3.75 to 15 C.c.).

DULCAMARA (Bittersweet), Woody Nightshade.

The young branches of Solanum Dulcamara (nat. ord. Solanaceæ), growing in the United States.

Contains an alkaloid Solanine and a glucoside dulcamarine, gummy extract, gluten, green wax, resin, benzoic acid, starch, lignin and various salts of lime.

Used chiefly in cutaneous eruptions. Official in the fluid extract.

EXTRACTUM DULCAMARÆ FLUIDUM (Fluid Extract of Dulcamara).

Prepared in the usual manner from Dulcamara No. 60, using dilute alcohol as the menstruum.

Dose: 30 to 60 minims (1.9 to 3.75 C.c.).

GELSEMIUM (Yellow Jasmine).

The rhizome and roots of Gelsemium sempervirens (nat. ord. Loganiaceæ), growing in the United States.

Contains an alkaloid Gelsemine, gelseminic acid, gum, starch, pectic acid, albumen, gallic acid, fixed oil, a fatty resin, a dry acrid resin, volatile oil, etc.

It is a cardiac sedative. Antidotes: Emetics and morphine, strychnine and atropine hypodermically. Official in the tincture and fluid extract.

EXTRACTUM GELSEMII FLUIDUM (Fluid Extract of Gelsemium).

Prepared in the usual manner from gelsemium No. 60, using alcohol as the menstruum.

Dose: 2 to 3 minims (0.12 to 0.18 C.c).

TINCTURA GELSEMII (Tincture of Gelsemium).

Prepared by macerating and percolating gelsemium No. 60 with a mixture of 650 parts alcohol and 350 parts water.

Dose: 10 to 20 minims (0.6 to 1.25 C.c.).

GRANATUM (Pomegranate).

The bark of the stem and root of Punica Granatum (nat. ord. Lythrarieæ), growing on the shores of the Mediterranean and in the East and West Indies.

Contains an alkaloid Pelletierine $C_8H_{13}NO$, upon which its properties depend, tannic acid, mannite, etc.

It is an astringent and vermifuge. Dose: 20 to 30 grains (1.3 to 1.95 gm.).

GRINDELIA (Grindelia).

The leaves and flowering tops of Grindelia robusta and of Grindelia squarrosa (nat. ord. Compositæ), growing in North and South America.

Contains an alkaloid Grindeline, an oil resembling oil of turpentine and resin. Official in the fluid extract. It is antispasmodic.

EXTRACTUM GRINDELIÆ FLUIDUM (Fluid Extract of Grindelia).

Prepared in the usual manner from grindelia No. 30, using alcohol as the menstruum. Dose: ½ to 1 fluid drachm (1.9 to 3.75 C.c.).

GUARANA (Guarana).

A dried paste chiefly consisting of the crushed or pounded seeds of Paullinia cupana (nat. ord. Sapindaceæ) growing in South America.

Contains an alkaloid Guaranine, identical with caffeine, tannic acid, gum, starch, albumen, a greenish fixed oil, etc.

It is a nerve stimulant. Official in the fluid extract.

EXTRACTUM GUARANÆ FLUIDUM (Fluid Extract of Guarana).

Prepared in the usual manner from guarana No. 80, using a mixture of 750 parts alcohol and 250 parts water as the menstruum.

Dose: 1 to 2 fluid drachms (3.75 to 7.5 C.c.).

CAFFEINA (Caffeine) $C_8H_{10}N_4O_2$, H_2O. Theine, Guaranine.

An alkaloid obtained from tea, coffee or guarana.

Permanent in the air. Soluble in water, alcohol and chloroform, sparingly soluble in ether.

Its aqueous solution should not be precipitated by mercuric potassium iodide test solution (absence of other alkaloids). It is a cerebral stimulant and diuretic. Dose: 3 to 8 grains (0.194 to 0.51 gm.).

CAFFEINA CITRATA (Citrated Caffeine) Citrate of Caffeine.

Prepared by dissolving caffeine in a solution of citric acid in water, evaporating to dryness, powdering and placing in well-stoppered bottles.

It is not regarded as a chemical salt, but a mechanical mixture. Soluble in water and alcohol. Used for the same purposes and in the same doses as caffeine.

CAFFEINA CITRATA EFFERVESCENS (Effervescent Citrate of Caffeine).

Contains 1% caffeine citrate, citric acid, sodium bicarbonate, tartartic acid and sugar. Dose: 1 teaspoonful.

HYDRASTIS (Golden Seal) Yellow Root, Yellow Puccoon, Orange Root, Indian Dye, Indian Tumeric.

The rhizome and roots of Hydrastis Canadensis (nat. ord. Ranunculaceæ) growing in the United States.

Contains two alkaloids Hydrastine $C_{21}H_{21}NO_6$ and Berberine, albumen, starch, fatty matter.

It is alterative, tonic, cholagogue, diuretic, etc. Official in the glycerite, fluid extract and tincture.

GLYCERITUM HYDRASTIS (Glycerite of Hydrastis).

Prepared by macerating and percolating hydrastis No. 60 with alcohol to exhaustion, adding water, distilling off the alcohol, filtering and adding glycerin.

Dose: ½ to 1 fluid drachm (1.87 to 3.75 C.c.).

EXTRACTUM HYDRASTIS FLUIDUM (Fluid Extract of Hydrastis).

Prepared in the usual manner from hydrastis No. 60, using a mixture of alcohol, glycerin and water as the menstruum.

Dose: ½ to 1 fluid drachm (1.87 to 3.75 C.c.).

TINCTURA HYDRASTIS (Tincture of Hydrastis).

Prepared by macerating and percolating hydrastis No. 60 with dilute alcohol.

Dose: ½ to 1 fluid drachm (1.87 to 3.75 C.c.).

HYDRASTININE HYDROCHLORAS (Hydrastinine Hydrochlorate) $C_{11}H_{11}NO_2HCl$.

The hydrochlorate of an artificial alkaloid derived from hydrastine, the latter being a colorless alkaloid obtained from hydrastis.

Deliquescent on exposure to damp air, and should be kept in well-stoppered vials. Soluble in water and alcohol; sparingly soluble in ether and chloroform.

It is an active oxytoxic. Dose: ¾ to 1½ grains (0.05 to 0.1 gm.).

HYOSCYAMUS (Henbane).

The leaves and flowering tops of Hyoscyamus niger (nat ord. Solanaceæ) collected from plants of the second year's growth.

Contains a crystallizable alkaloid Hysocyamine and an amorphous one Hyoscine. Similar in its action to belladonna, but is more of a hypnotic.

Official in the extract, fluid extract and tincture. Dose: 5 to 10 grains (0.33 to 0.65 gm.).

EXTRACTUM HYOSCYAMI (Extract of Hysocyamus).

Prepared by exhausting hyocyamus No. 60 with a mixture of alcohol and water, distilling off the alcohol, and evaporating the remainder to a pilular consistence.

Dose: 1 to 2 grains (0.065 to 0.13 gm.).

EXTRACTUM HYOSCYAMI FLUIDUM (Fluid Extract of Hyoscyamus).

Prepared in the usual manner from hyoscyamus No. 60, using a mixture of 2 parts alcohol and 1 part water as a menstruum.

Dose: 5 minims (0.3 C.c.).

TINCTURA HYOSCYAMI (Tincture of Hyoscyamus).

Prepared by macerating and percolating hyoscyamus No. 60 with dilute alcohol.

Dose: 1 fluid drachm (3.75 C.c.).

HYOSCINÆ HYDROBROMAS (Hyoscine Hydrobromate) $C_{17}H_{21}NO_4HBr, 3HO$. It should be kept in well-stoppered small vials.

Permanent in the air. Soluble in water and alcohol; sparingly soluble in ether and chloroform.

It is soporific. Dose: $\frac{1}{120}$ to $\frac{1}{60}$ of a grain (0.00054 to 0.0011 gm.).

HYOSCYAMINE HYDROBROMAS (Hyoscyamine Hydrobromate) $C_{17}H_{23}NO_3HBr$.

The hydrobromate of an alkaloid obtained from hyoscyamus, and should be kept in small, well-stoppered vials.

Deliquescent on exposure to the air. Soluble in water and alcohol. Its medical properties are the same as that of the sulphate.

HYOSCYAMINÆ SULPHAS (Hyoscyamine Sulphate) $(C_{17}H_{23}NO_3)_2 H_2SO_4$.

The neutral sulphate of an alkaloid obtained from hyoscyamus, and should be kept in small, well-stoppered vials.

Deliquescent in damp air. Soluble in water and alcohol. Sparingly soluble in ether and chloroform. It is a midriatic, but less powerful than atropine. Dose: $\frac{1}{80}$ of a grain (0.0008 gm.).

IPECACUANHA (Ipecac).

The root of Cephælis Ipecacuanha (nat. ord. Rubiaceæ) growing in Brazil.

Contains an alkaloid Emetine $C_{23}H_{40}N_2O_5$ (Wurtz), extractive, sugar, resin, volatile oil, and a glucoside ipecacuanhic acid.

Official in the fluid extract, powder of ipecac and opium, troches, troches of morphine and ipecac. It is emetic, diaphoretic and expectoraut.

EXTRACTUM IPECACUANHÆ FLUIDUM (Fluid Extract of Ipecac).

Prepared in the usual manner from ipecac No. 80, using a mixture of 750 parts alcohol and 250 parts water as the menstruum.

Official in tincture of ipecac and opium, syrup and wine.

Dose as an emetic: 15 to 30 minims (0.9 to 1.9 C.c.).

SYRUPUS IPECACUANHÆ (Syrup of Ipecac).

Prepared by adding water with acetic acid to fluid extract of ipecac, filtering, adding glycerin, and dissolving sugar in the solution. One fluid ounce contains the virtues of about 30 grains of ipecac. Dose as an expectorant: 30 to 60 minims (1.9 to 3.7 C.c.). As an emetic: 4 to 8 fluid drachms (15 to 30 C.c.).

VINUM IPECACUANÆ (Wine of Ipecac).

Prepared by adding fluid extract of ipecac to a mixture of alcohol and white wine, setting aside for a few days and filtering.

Dose as an emetic: 1 fluid ounce (30 C.c.). As an expectorant and diaphoretic: 10 to 30 minims (0.6 to 1.9 C.c.).

TROCHISCI IPECACUANHÆ (Troches of Ipecac).

Contains ipecac No. 60, tragacanth, sugar and syrup of orange. Each lozenge contains about one-third of a grain of ipecac.

LOBELIA (Lobelia) Indian Tobacco.

The leaves and tops of Lobelia inflata (nat. ord. Lobeliaceæ) collected after a portion of the capsules have been inflated. Growing in all parts of the United States.

Contains a volatile alkaloid Lobeline, volatile oil, lobelic acid, gum, resin, chlorophyll, fixed oil, lignin, etc.

Official in the fluid extract and tincture. Used principally in the treatment of spasmodic asthma.

EXTRACTUM LOBELIÆ FLUIDUM (Fluid Extract of Lobelia).

Prepared in the usual manner from Lobelia No. 60, using dilute alcohol as the menstruum.

Dose as an expectorant: 1 to 5 minims (0.06 to 0.3 C.c.). As an emetic: 10 to 20 minims (0.6 to 1.25 C.c.).

TINCTURA LOBELIÆ (Tincture of Lobelia).

Prepared by macerating and percolating lobelia No. 40 with dilute alcohol. Dose: ½ to 1 fluid drachm (1.9 to 3.75 C.c.).

MENISPERMUM (Canadian Moonseed) Yellow Parilla.

The rhizome and roots of Menispermum Canadense (nat. ord. Menispermaceæ) growing in the United States.

Contains an alkaloid Menispine, berberine, starch, etc.

Official in the fluid extract. It is a tonic.

EXTRACTUM MENISPERMI FLUIDUM (Fluid Extract of Menispermum).

Prepared in the usual manner from menispermum No. 60, using a mixture of 2 parts alcohol and 1 part water as the menstruum.

Dose: $\frac{1}{2}$ to 1 fluid drachm (1.9 to 3.75 C.c.).

NUX VOMICA (Nux Vomica) Quaker Buttons, Poison Nut.

The seed of Strychnos Nux Vomica (nat. ord. Loganiaceæ) growing in the East Indies Bengal, Ceylon, Malabar, and in many islands of the Indian Archipelago.

Contains two alkaloids Strychnine $C_{21}H_{22}N_2O_2$ and Brucine $C_{23}H_{26}N_2O_4$, yellow coloring matter, a concrete oil, gum, starch, wax, etc.

Official in the extract and fluid extract. It is a cardiac stimulant.

EXTRACTUM NUCIS VOMICÆ (Extract of Nux Vomica).

Prepared by macerating and percolating nux vomica No. 60 with a mixture of acetic acid, alcohol and water to exhaustion, distilling off the alcohol, shaking with ether, removing the ether, concentrating by evaporation, estimating the alkaloids, and adjusting to the strength of 15% of total alkaloids with sugar of milk and keeping in well-stoppered bottles.

The acetic acid is used to soften the tough tissues of the nux vomica, and the ether is used to extract the oily substances.

Official in the tincture. Dose: $\frac{1}{4}$ of a grain (0.016 gm.), equivalent to about $\frac{1}{27}$ of a grain of total alkaloids.

TINCTURA NUCIS VOMICÆ (Tincture of Nux Vomica.)

Prepared by dissolving extract of nux xomica, dried at a temperature of 100°C (212°F), in a mixture of 3 volumes alcohol and 1 volume water, and contains 2% extract, equivalent to 0.3% total alkaloids.

Dose: 20 minims (1.25 C.c.).

EXTRACTUM NUCIS VOMICÆ FLUIDUM (Fluid Extract of Nux Vomica).

Prepared by digesting and percolating nux vomica No 60 with a mixture of acetic acid, alcohol and water, to exhaustion, distilling off the alcohol, concentrating the remainder by evaporation, assaying and adjusting so that it will contain 1.5% of total alkaloids, with a mixture of 3 parts alcohol and 1 part water.

Dose: 3 minims (0.18 C.c.), equivalent to about $\frac{1}{23}$ of a grain of total alkaloids.

STRYCHNINA (Strychnine) $C_{21}H_{22}N_2O_2$.

An alkaloid obtained from nux vomica, and also obtained from plants of the Loganiaceæ.

Permanent in the air. Sparingly soluble in water, but soluble in alcohol and chloroform; almost insoluble in ether. It is liable to contain impurities, the chief of which, besides brucine, are coloring matter, lime and magnesia. Nitric acid does not redden it if it is pure, but reddens it if it contains brucine. Neither nitric acid or sulphuric acid color pure strychnine. Used in preparing citrate of iron and strychnine, and syrups of the phosphates of iron, quinine and strychnine,

LECTURE NO. 23.

STRYCHNINA (Continued).

It is a cardiac stimulant and tonic. There is no reliable chemical antidote to strychnine; tannic acid or soluble iodine, or a soluble iodide may be given which form alkaloidal compounds sparingly soluble, but strychnine iodide and tannate are poisonous and are capable of causing death. It is necessary first to produce vomiting, then absolute quiet, and the use of such spinal sedatives as chloroform, amyl nitrite, opium, chloral, and potassium bromide. The treatment of strychnine poisoning should be confined wholly to the physician and not attempted by the pharmacist. Dose the same as that of the sulphate.

STRYCHNINÆ SULPHAS (Strychnine Sulphate) $(C_{21}H_{22}N_2O_2)_2 H_2SO_4, 5H_2O$. It should be kept in well-stoppered vials.

Efflorescent in dry air. Soluble in alcohol and more so in water. Almost insoluble in ether.

Dose: $\frac{1}{20}$ of a grain (0.0032 gm.).

OPIUM (Opium).

The concrete, milky exudation obtained by incising the unripe capsules of Papaver somniferum (nat. ord. Papaveraceæ) and should yield in its normal, moist condition not less than 9% of crystallized morphine. It grows in India, Persia, Egypt and Asiatic Turkey.

Contains 19 alkaloids of which Morphine and Codeine are official; two acids are found combined with these alkaloids, namely: Meconic and Lactic acid, together with mucilage, pectic matter, and a glucose sugar.

Incompatible with alkalies, vegetable infusions containing tannic or gallic acids. It is a stimulant narcotic. Dose: 1 grain (0.065 gm.).

Antidotes: Emetics, stimulants and keep the patient in motion.

OPIUM PULVIS (Powdered Opium).

Prepared by drying opium at a temperature not exceeding 85°C(185°F), reducing to a fine No. 80 powder, and should contain not less than 13% nor more than 15% of crystallized morphine.

Official in the vinegar, extract, deodorized opium, pill, powder of ipecac and opium, tincture, deodorized tincture, troches of licorice and opium and wine.

OPIUM DEODORATUM (Deodorized Opium) Denarcotized Opium.

Prepared by macerating powdered opium with ether and decanting (to remove the odorous principles), repeating the maceration twice, drying the washed opium and adding sufficient sugar of milk to make the original weight.

ACETUM OPII (Vinegar of Opium), Black Drop.

Contains 10% of powdered opium, nutmeg, sugar and dilute acetic acid. Dose: 10 to 15 drops (0.6 to 1 C.c.).

EXTRACTUM OPII (Extract of Opium).

Prepared by triturating powdered opium with water, filtering, concentrating the filtrate, estimating the morphine, and adding sugar of milk so that the finished extract will contain 18% of morphine, and should be kept in well-stoppered vials.

Dose : ½ grain (0.033 gm.). Official in the plaster.

EMPLASTRUM OPII (Plaster of Opium).

Contains extract of opium 6%, burgundy pitch and lead plaster.

PILULÆ OPII (Pills of Opium).

Each pill contains 1 grain of powdered opium made into a mass with soap and water. Dose : 1 pill.

PULVIS IPECACUANHÆ ET OPII (Powder of Ipecac and Opium) Dover's Powder.

Contains 10% each of powdered ipecac and opium and 80% sugar of milk. It is an anodyne diaphoretic. Dose: 5 to 15 grains (0.33 to 1 gm.).

TINCTURA OPII (Tincture of Opium) Laudanum, Tincture of Thebaica.

Prepared by mixing together powdered opium and precipitated calcium phosphate, macerating with water for twelve hours, which has been heated to not over 90°C (194°F), adding alcohol, percolating and making up the loss with dilute alcohol.

Contains 10% of powdered opium, equivalent to from 1.3 to 1.5% of morphine. Opium in fine powder cannot be easily permeated, but if mixed with ½ its weight of an insoluble powder like calcium phosphate there will be no difficulty.

Dose: 11 minims or 22 drops (0.65 C.c.), equivalent to 1 grain of opium.

TINCTURA OPII CAMPHORATA (Camphorated Tincture of Opium) Paregoric, Compound Tincture of Camphor, Paregoric Elixir.

Contains 0.4% powdered opium, benzoic acid, camphor, oil of anise, glycerin and dilute alcohol.

Dose for an infant : 5 to 20 drops (0.3 to 1.25 C.c.). For an adult : 1 to 4 fluid drachms (3.75 to 15 C.c.).

TINCTURA OPII DEODORATA (Deodorized Tincture of Opium).

Prepared by mixing powdered opium with precipitated calcium phosphate, macerating with water for twelve hours, that has previously been heated to not over 90°C (194°F), percolating to exhaustion, concentrating by evaporation, shaking with ether, decanting off the ethereal layer and evaporating off the remainder, mixing the residue with water, filtering, and adding alcohol.

The strength is the same as that of tincture of opium. The ether is used to remove narcotine and other noxious matter.

Official in tincture of ipecac and opium. Its uses and doses are the same as those of tincture of opium.

TINCTURA IPECACUANHÆ ET OPII (Tincture of Ipecac and Opium) Tincture of Dover's Powder.

Prepared by evaporating deodorized tincture of opium and when cold adding fluid extract of ipecac, filtering and making up the required quantity with dilute alcohol.

Dose: 10 minims (0.6 C.c.), which represents 1 grain (0.065 gm.) each of powdered opium and ipecac.

TROCHISCI GLYCYRRHIZÆ ET OPII (Troches of Licorice and Opium) Opium Lozenges.

Contain extract of Licorice, powdered opium, acacia, sugar, oil of anise and water. Each troche contains about $\frac{1}{12}$ of a grain of opium.

VINUM OPII (Wine of Opium) Sydenham's Laudanum.

Prepared by percolating powdered opium, cassia cinnamon No.60, and cloves No. 60 mixed together with a mixture of white wine and alcohol.

The strength is the same as that of tincture of opium.

Dose: 15 to 20 drops (0.9 to 1.25 C.c.).

CODEINA (Codeine) $C_{18}H_{21}NO_3$, H_2O.

An alkaloid obtained from opium.

Efflorescent in the air. Soluble in water and more so in alcohol, ether and chloroform. It is extracted along with morphine in the preparation of the hydrochlorate; the solution when treated with ammonia precipitates the morphine, leaving the codeine in solution, which may be obtained by evaporation and crystallization.

It may be artificially prepared from morphine by treatment with methyl iodide and fixed alkali.

If codeine is added to nitric acid, and warmed, it turns yellow (difference from or absence of morphine). It is a very feeble alkaloid.

Dose: 1 grain (0.065 gm.).

MORPHINA (Morphine) $C_{17}H_{19}NO_3$, H_2O.

An alkaloid obtained from opium, by macerating with acid and water, precipitating with ammonia, purifying with alcohol, or by repeated solution in a dilute acid and precipitating. Permanent in the air. Sparingly soluble in water and more so in alcohol.

On treating morphine with cold concentrated sulphuric acid and adding a crystal of potassium permaganate, only a greenish color should be produced and no purple or violet color (difference from strychnine). On the addition of nitric acid it turns red (difference from quinine).

It is a narcotic. Dose: $\frac{1}{4}$ to $\frac{1}{2}$ grain (0.016 to 0.032 gm.).

MORPHINÆ ACETAS (Morphine Acetate) $C_{17}H_{19}NO_3.C_2H_4O_2$, $3H_2O$.

Prepared by dissolving morphine hydrochlorate in water, precipitating with solution of ammonia, dissolved in acetic acid, evaporating and crystallizing, and should be kept in well-stoppered, dark amber vials.

It slowly loses acetic acid when exposed to the air. Soluble in water and alcohol. Sparingly soluble in ether and chloroform. Its action and dose is the same as that of morphine.

MORPHINÆ HYDROCHLORAS (Morphine Hydrochlorate) $C_{17}H_{19}NO_3HCl$, $3H_2O$, Morphine Muriate.

Permanent in the air. Soluble in water and alcohol. Sparingly soluble in ether and chloroform.

Use and dose the same as that of morphine.

MORPHINÆ SULPHAS (Morphine Sulphate) $(C_{17}H_{19}NO_3)_2H_2SO_4$ $5H_2O$.

Prepared by dissolving morphine in sulphuric acid and water, kept hot, and setting aside to crystallize.

Permanent in the air. Soluble in water and less so in alcohol. Almost insoluble in ether. Its tests, uses and dose are the same as those for morphine.

Official in compound powder of morphine and troches of morphine and ipecac.

PULVIS MORPHINÆ COMPOSITUS (Compound Powder of Morphine) Tully's Powder.

Contains morphine sulphate, licorice No. 60, precipitated calcium carbonate, camphor, and should be kept in well-stoppered bottles.

Dose: 10 grains (0.65 gm.), containing $\frac{1}{6}$ of a grain (0.01 gm.) of morphine.

TROCHISCI MORPHINÆ ET IPECACUANHÆ (Troches of Morphine and Ipecac).

Contain morphine sulphate, ipecac No. 60, oil of wintergreen, sugar and mucilage of tragacanth.

Each troche contains $\frac{1}{40}$ of a grain of morphine (0.0016 gm.) and $\frac{1}{12}$ of a grain (0.005 gm.) of ipecac.

PAREIRA (Pareira Brava).

The root of Chondodendron tomentosum (nat. ord. Menispermaceæ), growing in Brazil. Contains an alkaloid Pelosine, a soft resin, etc.

Official in the fluid extract. It is tonic, aperient and diuretic.

EXTRACTUM PAREIRÆ FLUIDUM (Fluid Extract of Pareira).

Prepared in the usual manner from pareira No. 40, using a mixture of alcohol, glycerin and water as the menstruum.

Dose: 1 to 2 fluid drachms (3.75 to 7.5 C.c.).

PHYSOSTIGMA (Calabar Bean).

The seed of Physostigma venenosum (nat. ord. Leguminosæ), growing in Calabar, India and Brazil.

Contains an alkaloid Physostigmine.

Official in the extract and tincture. It is an alterative and myotic.

Antidotes: Emetics, stomach pump and atropine.

EXTRACTUM PHYSOSTIGMATIS (Extract of Physostigma).

Prepared by exhausting physostigma No. 80 with alcohol, distilling off the alcohol, and evaporating to a pilular consistence.

Dose: $\frac{1}{15}$ to $\frac{1}{6}$ of a grain (0.004 to 0.01 gm.).

TINCTURA PHYSOSTIGMATIS (Tincture of Physostigma).

Prepared by macerating and percolating physostigma No. 40 with alcohol.

Dose: 20 to 40 minims (1.25 to 2.5 C.c.).

PHYSOSTIGMINÆ SALICYLAS (Physostigmine Salicylate) $C_{15}H_{21}N_3O_2C_7H_6O_3$. Eserine Salicylate.

The salicylate of an alkaloid obtained from physostigma, and should be kept in small, dark amber colored, well-stoppered vials.

It acquires a reddish tint on exposure to light and air for a long time. Soluble in water and more so in alcohol.

It is a myotic. For use in the eyes a solution of 1 to 2 grains to the ounce is the usual strength. Dose internally: $\frac{1}{80}$ of a grain (0.0008 gm.).

PHYSOSTIGMINÆ SULPHAS (Physostigmine Sulphate) $(C_{15}H_{21}N_3O_2)_2H_2SO_4$. Eserine Sulphate.

The sulphate of an alkaloid obtained from physostigma, and should be kept in small, dark amber colored and well-stoppered vials.

It is very deliquescent when exposed to moist air, and gradually turns reddish when exposed to the air and light. Soluble in water and alcohol.

Its uses and doses are the same as that of the salicylate.

Eserine either in solution or in the form of gelatin disks containing from $\frac{1}{3000}$ to $\frac{1}{1000}$ of a grain, is used as a myotic.

PILOCARPUS (Jaborandi).

The leaflets of Pilocarpus Selloanus and of Pilocarpus Jaborandi, (nat. ord. Rutaceæ), growing in Brazil.

Contains two alkaloids Pilocarpine and Jaborine, tannic acid, a volatile acid, etc. Official in the extract. It is a diaphoretic.

Dose: 20 to 60 grains (1.3 to 3.9 gm.).

EXTRACTUM PILOCARPI FLUIDUM (Fluid Extract of Pilocarpus).

Prepared in the usual manner from pilocarpus No. 40, using dilute alcohol as the menstruum.

Dose: 15 to 30 minims (0.9 to 1.9 C.c.).

PILOCARPINÆ HYDROCHLORAS (Pilocarpine Hydrochlorate) $C_{11}H_{16}N_2O_2HCl$.

The hydrochlorate of an alkaloid obtained from pilocarpus, and should be kept in small well-stoppered vials.

Deliquescent on exposure to damp air. Soluble in water and alcohol. Almost insoluble in ether and chloroform. It is a diaphoretic and myotic. Dose: $\frac{1}{8}$ of a grain (.008 gm.).

PYRETHRUM (Pellitory) Pellitory of Spain.

The root of Anacyclus Pyrethrum (nat. ord. Compositæ), growing in the Levant, Barbary, and on the Mediterranean coast of Europe.

Contains an alkaloid Pyrethrine, fixed oil, tannin, gum, etc.

Official in the tincture. It is a powerful irritant and a Sialagogue.

TINCTURA PYRETHRI (Tincture of Pyrethrum).

Prepared by macerating and percolating pellitory No. 40 with alcohol.

It is a powerful local irritant, and is used in some mouth and tooth washes.

SANGUINARIA (Bloodroot) Puccoon, Tetterwort, Indian Paint.

The rhizome of Sanguinaria Canadensis (nat. ord. Papaveraceæ), growing in the United States.

Contains four alkaloids Sanguinarine, $C_{20}H_{15}NO_4$ chelerythrine, homochelidonine and protopine. It is a stimulant narcotic and expectorant. Official in the fluid extract and tincture. Dose as an emetic: 10 to 20 grains (0.65 to 1.3 gm.). As an expectorant: 1 to 5 grains (0.065 to 0.33 gm.).

EXTRACTUM SANGUINARIÆ FLUIDUM (Fluid Extract of Sanguinaria).

Prepared in the usual manner from sanguinaria No. 60, using a mixture of acetic acid, alcohol and water as the menstruum.

Dose: 3 to 5 minims (0.18 to 0.3 C.c.).

TINCTURA SANGUINARIÆ (Tincture of Sanguinaria).

Prepared by macerating and percolating bloodroot No. 60 with a mixture of acetic acid, alcohol and water as the menstruum.

The acetic acid in bloodroot preparations is to partially prevent precipitation. Dose: 30 to 60 drops (1.9 to 3.75 C.c.).

SCOPARIUS (Broom) Broom Tops

The tops of Cytisus Scoparius (nat. ord. Leguminoseæ), growing in the United States.

Contains an alkaloid called Sparteine $C_{15}H_{26}N_2$, a principle called Scoparin $C_{21}H_{22}O_{10}$, volatile oil, wax, fatty matter, chlorophyll, yellow coloring matter, tannin, mucilage, albumen, etc.

Official in the fluid extract. It is diuretic and cathartic.

EXTRACTUM SCOPARII FLUIDUM (Fluid Extract of Scoparius).

Prepared in the usual manner from scoparius No. 60, using dilute alcohol as a menstruum. Dose: 20 to 40 minims (1.3 to 2.6 C.c.).

SPARTEINÆ SULPHAS (Sparteine Sulphate) $C_{15}H_{26}N_2H_2SO_4$, $4H_2O$.

The neutral sulphate of an alkaloid obtained from scoparius.

It is liable to attract moisture when exposed to damp air. Soluble in alcohol and water.

It is a cardiac stimulant. Dose: $\frac{1}{6}$ of a grain (0.011 gm.).

SPIGELIA (Pinkroot).

The rhizome and rootlets of Spigelia marilandica (nat. ord. Leguminoseæ), growing in the United States.

Contains an alkaloid Spigeline, volatile oil, tannic acid, inert extractive, wax, lignin, resin, etc.

Official in the fluid extract. It is an anthelmintic.

EXTRACTUM SPIGELIÆ FLUIDUM (Fluid Extract of Spigelia).

Prepared in the usual manner from spigelia No. 60, using dilute alcohol as the menstruum.

Dose for an adult: 1 to 2 fluid drachms (3.75 to 7.5 C.c.). For children: 10 to 20 minims (0 6 to 1.25 C.c.).

STAPHISAGRIA (Stavesacre) Stavesacre Seed.

The seed of Delphinium Staphisagria (nat. ord. Ranunculaceæ), growing in the South of Europe.

Contains four alkaloids, Delphinine, $C_{28}H_{37}NO_5$, delphinoidine $C_{42}H_{68}N_2O_7$, delphisine $C_{27}H_{46}N_2O_4$, staphisagrine $C_{32}H_{35}NO_5$.

It is emetic and cathartic.

STILLINGIA (Queensroot) Silver Leaf, Queen's Delight.

The root of Stillingia sylvatica (nat. ord. Euphorbiaceæ), growing in United States.

Contains an alkaloid Stillingine and a volatile oil.

Official in the fluid extract. It is emetic, cathartic, and alterative.

EXTRACTUM STILLINGIÆ FLUIDUM (Fluid Extract of Stillingia).

Prepared in the usual manner from Stillingia No. 40, using dilute alcohol as the menstruum. Dose: 15 to 45 minims (0.9 to 2.8 C.c.)

STRAMONII FOLIA (Stramonium Leaves) Thornapple Leaves.

The leaves of Datura Stramonium (nat. ord. Solanaceæ) growing in Russia and the United States.

STRAMONII SEMEN (Stramonium Seed) Thornapple.

The seed of Datura Stramonium (nat. ord. Solanaceæ). Stramonium contains an alkaloid Daturine (which is isomeric with hyoscyamine) and atropine.

Official in the extract, fluid extract and tincture. Its uses are the same as those of belladonna.

EXTRACTUM STRAMONII SEMINIS (Extract of Stramonium Seed).

Prepared by macerating and percolating stramonium seed No. 60 with dilute alcohol, and evaporating to a pilular consistence. Dose: ⅛ to ¼ of a grain (0.010 to 0.016 gm.). Official in the ointment.

UNGUENTUM STRAMONII (Ointment of Stramonium).

Prepared by rubbing extract of stramonium seed with a little diluted alcohol, and mixed with benzoinated lard.

Used as an anodyne in ulcers, etc.

EXTRACTUM STRAMONII SEMINIS FLUIDUM (Fluid Extract of Stramonium Seed).

Prepared in the usual manner from stramonium seed No. 60, using a mixture of 750 parts alcohol and 250 water as the menstruum.

Dose: 1 to 2 minims (0.06 to 0 12 C.c.).

TINCTURA STRAMONII SEMINIS (Tincture of Stramonium Seed).

Prepared by macerating and percolating stramonium seed No. 40 with dilute alcohol. Dose: 20 to 30 minims (1.25 to 1.9 C.c.).

TABACUM (Tobacco) Leaf Tobacco.

The commercial, dried leaves of Nicotiana Tabacum (nat ord. Solanaceæ) growing in the United States.

Contains an alkaloid Nicotine $C_{10}H_{14}N_2$, a substance called nicotianin, bitter extractive, gum, green resin, albumen, etc.

It is a powerful sedative.

Nicotine is a powerful poison and exists in tobacco in small proportions. When cigars are smoked certain gases are given off consisting of oxygen, nitrogen, carbonic acid and marsh gas. Tobacco undergoes considerable chemical change during the process of curing and preparing for use. Prof. Mayer of New York has determined experimentally that nicotine exists as largely in the plant before as after curing. When employed in excess it impairs digestion, produces emaciation, general debility, and lays the foundation of serious nervous disorders.

VERATRUM VIRIDE (American Hellebore) Green Hellebore.

The rhizome and roots of Veratum viride (nat. ord. Liliaceæ) growing in the United States.

· Contains two alkaloids, Viridine and Veratroidine.

Official in the fluid extract and tincture. It is a depresso-motor. Antidotes: Whiskey or brandy internally, strychnine and digitalis hypodermically.

EXTRACTUM VERATRI VIRIDIS FLUIDUM (Fluid Extract of Veratrum Viride.

Prepared in the usual manner from veratrum viride No. 60, using alcohol as the menstruum.

Dose: 1 to 2 minims (0.06 to 0.12 C.c.).

TINCTURA VERATRI VIRIDIS (Tincture of Veratrum Viride).

Prepared by macerating and percolating veratrum viride No. 60 with alcohol. Dose: 3 to 8 minims (0.18 to 0.5 C.c.).

VERATRINA (Veratrine).

A mixture of alkaloids obtained from the seed of Asagræa officinalis (nat. ord. Liliaceæ).

Permanent in the air. Insoluble in water, but soluble in alcohol, ether and chloroform. Official in the ointment and oleate. It is irritant. Dose: $\frac{1}{30}$ of a grain (0.002 gm.).

LECTURE NO. 24.

OLEATUM VERATRINÆ (Oleate of Veratrine).
Contains 2% of veratine dissolved in oleic acid.
UNGUENTUM VERATRINÆ (Ointment of Veratrine).
Contains 4% of veratrine, mixed with olive oil and benzoinated lard.

RESINS.

CANNABIS INDICA (Indian Hemp) Hemp.
The flowering tops of the female plant of Cannabis Sativa (nat. ord. Urticaceæ), growing in India and Persia.

Contains a resin Cannabin, and volatile oil in small proportion.

Official in the extract, fluid extract, and tincture. It is a powerful narcotic.

EXTRACTUM CANNABIS INDICÆ (Extract of Cannabis Indica).

Prepared by macerating and percolating cannabis indica No. 20 with alcohol and evaporating to a pilular consistence.

As it varies in strength the commencing dose should be not over $\frac{1}{4}$ grain (0.016 gm.).

EXTRACTUM CANNABIS INDICÆ FLUIDUM (Fluid Extract of Cannabis Indica).

Prepared in the usual manner from cannabis indica No. 20, using alcohol as the menstruum. Dose: $\frac{1}{2}$ to 1 minim (0.03 to 0.06 C.c.).

TINCTURA CANNABIS INDICÆ (Tincture of Cannabis Indica).
Prepared by macerating and percolating cannabis indica No. 40 with alcohol. Dose: 30 minims (1.9 C c.).

CIMICIFUGA (Black Snakeroot) Black Cohosh.
The rhizome and roots of Cimicifuga racemosa (nat. ord. Ranunculaceæ) growing in the United States from Canada to Florida.

Contains a resinous amorphous body, gum, starch, sugar, wax, fatty matter, tannin, gallic acid, etc.

Official in the extract, fluid extract and tincture. Used principally in St. Vitus dance.

EXTRACTUM CIMICIFUGÆ (Extract of Cimicifuga).
Prepared by macerating and percolating cimicifuga No. 60 to exhaustion with alcohol, and evaporating to a pilular consistence.

Dose: 3 to 10 grains (0.194 to 0.65 gm.).

EXTRACTUM CIMICIFUGÆ FLUIDUM (Fluid Extract of Cimicifuga).

Prepared in the usual manner from cimicifuga No. 60, using alcohol as the menstruum. Dose: 30 to 60 minims (1.9 to 3.75 C.c.).

TINCTURA CIMICIFUGÆ (Tincture of Cimicifuga).
Prepared by macerating and percolating cimicifuga No.60 with alcohol.
Dose: 1 to 4 fluid drachms (3.75 to 15 C.c.).

RESINA COPAIBÆ (Resin of Copaiba).

The residue left after distilling off the volatile oil from copaiba, and contains copaivic acid $C_{20}H_{30}O_2$.

Dose: 10 to 20 grains (0.65 to 1.3 gm.).

CUSSO (Kousso) Brayera, Cousso, Kusso.

The female inflorescence of Hagenia Abyssinica (nat. ord. Rosaceæ) growing in Abyssinia.

Contains a bitter acrid resin Kosin, fatty matter, chlorophyll, wax, sugar, gum, tannic acid, etc.

Official in the fluid extract. It is a vermifuge.

EXTRACTUM CUSSO FLUIDUM (Fluid Extract of Kousso).

Prepared in the usual manner from kousso No. 40, using alcohol as the menstruum. Dose: $\frac{1}{2}$ to 1 fluid ounce (15 to 30 C.c.).

ERIODICTYON (Yerba Santa) Consumptives' Weed, Bears' Weed, Mountain Balm.

The leaves of Eriodictyon glutinosum (nat. ord. Hydrophyllaceæ), growing in California. Contains a bitter, acrid resin (upon which its activity depends), tannic acid, volatile oil, etc.

Official in the fluid extract. It is a bitter tonic and stimulant balsamic expectorant.

EXTRACTUM ERIODICTYI FLUIDUM (Fluid Extract of Eriodictyon).

Prepared in the usual manner from eriodictyon No. 60, using a mixture of 800 parts alcohol and 200 water as the menstruum.

Dose: 20 to 60 minims (1.3 to 3.75 C.c.).

ELASTICA (India Rubber) Caoutchouc.

The prepared milk juice of various species of Hevea (nat. ord. Euphorbiaceæ) known in commerce as Para Rubber, and grows in South America.

Insoluble in water, dilute acids or dilute solutions of alkalies. Soluble in chloroform, carbon disulphide, oil of turpentine, benzin or benzol.

GOSSYPII RADICIS CORTEX (Cotton Root Bark).

The bark of the root of Gossypium herbaceum, and of other species of Gossypium (nat. ord. Malvaceæ), growing in Asia, Africa and the United States.

Contains a peculiar acrid resin upon which its properties depend.

Official in the fluid extract.

EXTRACTUM GOSSYPII RADICIS FLUIDUM (Fluid Extract of Cotton Root Bark).

Prepared in the usual manner from cotton root bark No. 30, using a mixture of alcohol and glycerin as the menstruum.

The glycerin is used to prevent gelatinization. Dose: $\frac{1}{2}$ to 1 fluid drachm (1.9 to 3.75 C.c.).

GUIACI LIGNUM (Guiacum Wood) Lignum Vitæ.

The heart wood of Guiacum officinale and of Guaiacum sanctum (nat. ord. Zygophylleæ) growing in the West Indies.

Contains a resin, bitter pungent extractive, etc.

Official in compound decoction of sarsaparilla. It is a stimulant diaphoretic.

GUAIACI RESINA (Guaiac).

The resin of the wood of Guaiacum officinalis, obtained by natural exudation, by incision or by heat.

Exposed to the air and light it absorbs oxygen and becomes green, and the change takes place rapidly in sunshine. Soluble in alcohol, ether and alkaline solutions. Often adulterated with the resin of the pine.

Official in the tincture, ammoniated tincture and compound pill of antimony. It is a stimulant and alterative.

TINCTURA GUAIACI (Tincture of Guaiac).

Prepared by macerating guaiac in coarse powder in alcohol for seven days and percolating.

Used in cases of gout and rheumatism. Dose: 1 teaspoonful (3.7 C.c).

TINCTURA GUAIACI AMMONIATA (Ammoniated Tincture of Guaiac).

Prepared by macerating guaiac in coarse powder, in aromatic spirit of ammonia for seven days, and filtering.

Used in cases of chronic rheumatism. Dose: 1 to 2 fluid drachms (3.75 to 7.5 C.c.).

JALAPA (Jalap).

The tuberous root of Ipomœa Jalapa (nat. ord. Convolvulaceæ), growing in Europe, India and Mexico.

Contains a resin Jalapin, gummy extract, albumen, etc.

Official in the alcoholic extract and compound powder.

It is a cathartic. Dose: 15 to 30 grains (0.97 to 1.95 gm.).

EXTRACTUM JALAPÆ (Extract of Jalap).

Prepared by macerating and percolating jalap No. 60 with alcohol and evaporating to a pilular consistence.

Official in compound cathartic pill and vegetable cathartic pill.

Dose: 10 to 20 grains (0.65 to 1.3 gm.).

PULVIS JALAPÆ COMPOSITUS (Compound Powder of Jalap).

Contains 35% of jalap No. 60, and 65% of potassium bitartrate.

Dose: 30 to 60 grains (1.95 to 3.9 gm.).

RESINA JALAPÆ (Resin of Jalap).

Prepared by macerating and percolating jalap No. 60 with alcohol, distilling off alcohol, reducing to small bulk, and adding to water.

Often adulterated with guaiac resin, etc. Dose: 2 to 5 grains (0.13 to 0.33 gm.).

MASTICHE (Mastic).

A concrete resinous exudation from Pistacia Lentiscus (nat. ord. Anacardieæ) obtained by making incisions in the bark of the stem and large branches. It grows in the island of Scio in the Grecian Archipelago.

Contains a Resin, masticin and a volatile oil. Its properties are analagous to those of the turpentines. Official in pill of aloes and mastic.

PEPO (Pumpkin Seed).

The seed of Cucurbita Pepo (nat. ord. Cucurbitaceæ) growing all over the world.

Contains a Resin, fixed oil, an aromatic principle, chlorophyll, starch, sugar, and an alkaloid cucurbitine.

Used as a vermifuge.

PIX BURGUNDICA (Burgundy Pitch).

The prepared resinous exudation of Abies excelsa (nat. ord. Coniferæ), growing in Europe and Northern Asia.

Almost entirely soluble in glaciale acetic acid, in boiling alcohol and partially soluble in cold alcohol.

Official in the plaster, and pitch plaster with cantharides.

It is a gentle rubifacient.

EMPLASTRUM PICIS BURGUNDICÆ (Burgundy Pitch Plaster).

Contains Burgundy pitch, olive oil and yellow wax.

EMPLASTRUM PICIS CANTHARIDATUM (Cantharidal Pitch Plaster) Warming Plaster.

Contains cerate of cantharides and Burgundy pitch.

KAMALA (Kamala) Rottlera, Spoonweed.

The glands and hairs from the capsules of Mallotus Philippinensis (nat. ord. Euphorbiaceæ) growing in Hindoostan.

Contains a resinous coloring substance Rottlerin, albumen, cellulose, volatile oil, etc.

It is a vermifuge. Dose: 1 to 3 drachms (3.9 to 11.65 gm.).

PODOPHYLLUM (May-Apple) Mandrake Root.

The rhizome and roots of Podophyllum peltatum (nat ord. Berberideæ) growing in the United States and Japan.

Contains a resin Podophyllin, albumen, gum, starch, extractive, gallic acid, fixed oil, traces of volatile oil, etc.

Official in the extract and fluid extract. It is a cathartic.

EXTRACTUM PODOPHYLLI (Extract of Podophyllum).

Prepared by macerating and percolating podophyllum No. 60 with a mixture of 800 parts alcohol and 200 parts water, to exhaustion and evaporating to a pilular consistence.

Dose: 1 to 3 grains (0.065 to 0.20 gm.).

EXTRACTUM PODOPHYLLI FLUIDUM (Fluid Extract of Podophyllum).

Prepared in the usual manner from podophyllum No. 60, using a mixture of 800 parts alcohol and 200 parts water as the menstruum.

Dose: 5 to 15 minims (0.3 to 0.9 C.c.).

RESINA PODOPHYLLI (Resin of Podophyllum) Resin of May-Apple.

Prepared by macerating and percolating podophyllum No. 60 with alcohol, distilling off the alcohol until a syrupy consistence is obtained, and precipitating with water to which hydrochloric acid has been added. The acid aids in the precipitation of the resin. It consists of two resins, one soluble in ether and alcohol, the other in alcohol only.

It is a powerful cathartic. Dose: $\frac{1}{8}$ to $\frac{1}{2}$ grain (0.008 to 0.03 gm.).

RESINA (Resin) Colophony.

The residue left after distilling off the volatile oil from turpentine.

Soluble in alcohol, ether, fixed and volatile oils.

White Rosin differs from resin only in being opaque and of a whitish color, owing to the water with which it is incorporated, and which gradually escapes on exposure, leaving it more or less transparent.

Official in the cerate and plaster.

CERATUM RESINÆ (Resin Cerate) Basilicon Ointment.

Contains resin, yellow wax and lard.

Used as a stimulant application to burns and blistered surface.

EMPLASTRUM RESINÆ (Resin Plaster) Adhesive Plaster.

Contains resin, lead plaster and yellow wax.

Used for retaining the sides of wounds together.

RHAMNUS PURSHIANA (Cascara Sagrada) Sacred Bark, Chittem Bark.

The bark of Rhamnus Purshiana (nat. ord. Rbamnaceæ) growing in California. Contains a bitter resin, a red resin, light yellow resin, tannic, oxalic and malic acids, a neutral crystallizable substance, volatile oil, a ferment, glucose and ammonia.

According to Meier and Webber the ferment causes griping, and by keeping the bark at least two years before being used, the ferment has exhausted itself, and this objection is overcome. The laxative effect is due to the resins and the tonic effects to the crystalline bitter principle.

It is a laxative and tonic. Official in the fluid extract.

EXTRACTUM RHAMNI PURSHIANÆ (Fluid Extract of Cascara Sagrada).

Prepared in the usual manner from cascara sagrada No. 60, using dilute alcohol as the menstruum. Dose: 15 to 45 minims (0.9 to 2.7 C.c.).

SCAMMONIUM (Scammony).

A resinous exudation from the living root of Convolvulus Scammonia (nat. ord. Convolvulaceæ) growing in Syria, Anatolia and certain islands of the Archipelago.

Contains a resin, gum, etc. It is a cathartic.

RESINA SCAMMONII (Resin of Scammony).

Prepared by digesting scammony No. 60 with alcohol in successive portions until exhausted, mixing the tincture, and precipitating with water. Often adulterated with resin of jalap, rosin, guaiac, etc.

Dose: 4 to 8 grains (0.26 to 0.52 gm.).

Gum Resins.

AMMONIACUM (Ammoniac).

A gum resin obtained from Dorema Ammoniacum nat. ord. Umbelliferæ) exuding from the stem, and grows in the Persian Provinces.

Contains gum, resin, a gluten-like substance (bassorin), volatile oil, etc.

It comes either in the state of tears, or in aggregated masses, and in both forms is frequently mixed with impurities. The U. S. P. 1890 requires that only such masses as are composed of tears should be considered up to the standard. When triturated with water it forms an emulsion. Official in ammoniac plaster with mercury, and the emulsion. It is a stimulant and expectorant.

Dose: 10 to 30 grains (0.65 to 1.30 gm.).

EMULSUM AMMONIAICI (Emulsion of Ammoniac) Ammoniac Mixture, Milk of Ammoniac.

Prepared by rubbing ammoniac in a warm mortar with water and straining.

The gum is dissolved in the water and holds the resin in suspension.

Dose: 1 to 2 tablespoonfuls (15 to 30 C.c.)

ASAFŒTIDA (Asafetida).

A gum resin obtained from the root of Ferula fœtida (nat. ord. Umbelliferæ) obtained by incision, and grows in Persia.

Contains gum, resin, bassorin, volatile oil, etc.

The oldest plants are the most productive. On keeping for a long time exposed, it hardens and becomes brittle and the intensity of its smell and taste diminishes. When triturated with water it forms an emulsion. At least 60% of it should dissolve in alcohol.

Often adulterated with inferior qualities, sand and stones. It is often kept in a powdered state, but this is unfit for use on account of the loss of volatile oil and its liability to adulteration.

Official in pills, pill of aloes and asafœtida, emulsion and tincture. It is a stimulant antispasmodic, expectorant, and laxative.

Dose: 10 grains (0.65 gm.).

EMULSUM ASAFŒTIDÆ (Emulsion of Asafetida) Asafetida Mixture, Milk of Asafetida.

Prepared by rubbing asafetida in a warm mortar with water, and contains 4% of asafetida. Dose: 4 to 8 fluid drachms (15 to 30 C.c.).

PILULÆ ASAFŒTIDÆ (Pills of Asafetida).

Each pill contains 3 grains of asafetida made into a mass with soap and water.

TINCTURA ASAFŒTIDÆ (Tincture of Asafetida).

Prepared by macerating asafetida (bruised) in alcohol for 7 days, and filtering. Dose: 20 to 30 minims (1.25 to 1.9 C.c.).

GAMBOGIA (Gamboge).

A gum resin obtained from Garcinia Hanburii (nat. ord. Guttiferæ), growing in Siam and Cochin-China.

Partly soluble in alcohol and ether. When triturated with water it forms an emulsion. It is a gum resin, without volatile oil.

Official in compound cathartic pill. It is a powerful hydragogue cathartic. Dose: 2 to 6 grains (0.13 to 0.4 gm.).

MYRRHA (Myrrh).

A gum resin obtained from Commiphora Myrrha(nat.ord.Burseraceæ), growing in Arabia and Africa. Contains gum, resin and volatile oil.

Soluble in solutions of alkalies. Partially soluble in water, alcohol and ether. When triturated with water it forms an emulsion. Often adulterated with pieces of bdellium and many other gummy or resinous substances.

Official in compound iron mixture, pill of aloes and myrrh, tincture of aloes and myrrh, and tincture.

It is a stimulant tonic. Dose: 10 to 30 grains (0.65 to 1.95 gm.).

TINCTURA MYRRHÆ (Tincture of Myrrh).

Prepared by macerating myrrh in moderately coarse powder in alcohol for 7 days and filtering.

Used as a local application, and as a stimulant expectorant and emmenagogue. Dose: 15 to 30 minims (0.9 to 1.9 C.c.).

OLEORESINS.

ASPIDIUM (Male Fern) Male Shield Fern.

The rhizome of Dryopteris Felix-mas and Dryopteris marginalis (nat. ord. Felices), growing in Europe, Asia and Africa.

Contains volatile oil, resin and filicic acid. Official in the oleoresin. It is a vermifuge.

OLEORESINA ASPIDII (Oleoresin of Aspidium) Liquid Extract of Male Fern, Oil of Fern.

Prepared by percolating male fern with ether, evaporating off the ether, and should be kept in well stoppered bottles.

It usually deposits on standing, a granular crystalline precipitate, which should be thoroughly mixed with the liquid portion before being used. Dose: ½ to 1 fluid drachm (1.9 to 3.75).

CAPSICUM (Capsicum) Cayenne Pepper, African Pepper.

The fruit of Capsicum fastigiatum (nat. ord. Solanaceæ), growing in most all parts of the world. Contains a volatile oil, resin and capsaicin. Sometime adulterated with colored sawdust which may be recognized by the microscope. Official in the fluid extract and tincture. It is a local stimulant. Dose: 5 to 10 grains (0.33 to 0.65 gm.).

EXTRACTUM CAPSICI FLUIDUM (Fluid Extract of Capsicum).

Prepared by macerating and percolating capsicum No. 60 with alcohol in the usual manner. Dose: ½ to 1 minim (0.03 to 0.06 C.c.).

OLEORESINA CAPSICI (Oleoresin of Capsicum).

Prepared by percolating capsicum No. 60 with ether and recovering the ether, and should be kept in well-stoppered bottles.

Official in capsicum plaster. Dose: ¼ to 1 minim (0.015 to 0.06 C.c.).

EMPLASTRUM CAPSICI (Capsicum Plaster).

Contains oleoresin of capsicum and resin plaster.

It is a rubifacient.

TINCTURA CAPSICI (Tincture of Capsicum).

Prepared by macerating and percolating capsicum No. 30 with a mixture of 950 parts alcohol and 50 parts water.

Dose: ½ to 1 fluid drachm (1.9 to 3.75 C.c.).

COPAIBA (Copaiba) Balsam of Copaiba.

The oleoresin of Copaiba Langsdorffi (nat. ord. Leguminoseæ), obtained by cutting deeply or boring into the trunk of the tree. It grows in Brazil and other parts of South America. Contains volatile oil, resin and copaivic acid.

The Para balsam is the lighter color and the best quality. Insoluble in water, soluble in alcohol, ether, fixed and volatile oils, chloroform, carbon disulphide and benzin. On exposure to the air it acquires a deeper color, a thicker consistence, and greater density, and, if spread out upon an extended surface, ultimately becomes dry and brittle. Often adulterated with Gurjan Balsam, turpentine and fixed oils.

Official in the mass. It is a stimulant, diuretic, laxative, and in large doses purgative. Dose: 20 minims to 1 fluid drachm (1.25 to 3.75 C.c.).

MASSA COPAIBÆ (Mass of Copaiba) Solidified Copaiba.

Prepared by triturating magnesia with water, adding balsam of copaiba, heating on a water bath for half an hour and setting aside until it has acquired a pilular consistence.

The magnesia combines chemically with the copaivic acid of copaiba, and acts as an absorbent to the volatile oil. It is convenient for the preparation of pills. Dose: 5 grains (0.33 gm.).

HUMULUS (Hops).

The strobiles of Humulus Lupulus (nat. ord. Urticaceæ) growing in Europe and the United States.

Contains a principle called Lupulin upon which its properties depend.

Official in the tincture. It is a tonic and slightly narcotic.

TINCTURA HUMULI (Tincture of Hops).

Prepared by macerating and percolating hops well dried No. 20, with dilute alcohol.

Dose: 1 to 3 fluid drachms (3.7 to 11.25 C.c.).

LECTURE NO. 25.

LUPULINUM (Lupulin).

The glandular powder separated from the strobiles of Humulus Lupulus. Contains resinous matter and volatile oil (Valerol $C_6H_{10}O$).

Official in the fluid extract and oleoresin. It is tonic and slightly narcotic. Dose: 5 to 12 grains (0.325 to 0.78 gm.).

EXTRACTUM LUPULINI FLUIDUM (Fluid Extract of Lupulin).

Prepared in the usual manner from lupulin, using alcohol as the menstruum. Owing to its resinous character it is not miscible with aqueous fluids. Dose: 10 to 15 minims (0.6 to 0.9 C.c.).

OLEORESINA LUPULINI (Oleoresin of Lupulin).

Prepared by percolating lupulin with ether, evaporating off the ether, and preserving in close-stoppered bottles.

Dose: 2 to 5 grains (0.13 to 0.33 gm.).

PIPER (Black Pepper).

The unripe fruit of Piper nigrum (nat. ord. Piperaceæ), growing on the coast of Malabar, in the peninsula of Molucca, in Siam, Sumatra, Java, Borneo, Phillipines and West Indies. Contains Volatile Oil, resin and piperine.

Official in the oleoresin. It is a stimulant.

OLEORESINA PIPERIS (Oleoresin of Pepper) Fluid Extract of Black Pepper.

Prepared by percolating black pepper No. 60 with ether, recovering the ether and preserving in well-stoppered bottles.

Dose: $\frac{1}{4}$ to 1 minim (0.015 to 0.06 C.c.).

PIX LIQUIDA (Tar).

An empyreumatic oleoresin obtained by the destructive distillation of the wood of Pinus palustris, and of other species of Pinus (nat. ord. Coniferæ) growing in all parts of the United States.

Consists of resinous matter, united with acetic acid, oil of turpentine, and various empyreumatic products and colored with charcoal. By age it becomes granular and opaque. By distillation it yields pyroligneous acid and oil of tar. What is left is called pitch.

Official in the syrup and ointment. Its medical properties are the same as those of the turpentines, only less irritant.

Dose: $\frac{1}{2}$ to 1 drachm (1.9 to 3.75 C.c.).

SYRUPUS PICIS LIQUIDÆ (Syrup of Tar).

Prepared by mixing tar with washed white sand, adding water, macerating 12 hours, throwing the water away, pouring boiling water over the residue, adding glycerin, setting aside for 24 hours, decanting the clear solution, filtering and dissolving sugar in the filtrate.

The object of washing the tar is to remove the acid constituents which would be irritating. Dose: 1 to 2 fluid drachms (3.7 to 7.5 C.c.).

UNGUENTUM PICIS LIQUIDÆ (Tar Ointment).

Prepared by melting together lard and yellow wax, then incorporating 50% of tar with it.

Used as a stimulant application to scabby ulcers and sores.

SUMBUL (Sumbul).

The root of Ferula Sumbul (nat. ord. Umbelliferæ), growing in India.

Contains volatile oil, two balsamic resins, wax, gum, starch, sumbulic acid, and a little valerianic acid.

Official in the tincture. It is a nerve stimulant.

TINCTURA SUMBUL (Tincture of Sumbul).

Prepared by macerating and percolating sumbul No. 30 with a mixture of 650 parts alcohol and 350 parts water.

Dose: 20 to 60 minims (1.2 to 3.75 C.c.).

TEREBINTHINA (Turpentine) Crude Turpentine, Common Frankincense.

A concrete oleoresin obtained from Pinus palustris (nat. ord. Coniferæ) by scraping off the trunks of the tree. Contains resin associated with volatile oil.

The term Turpentine is usually applied to certain vegetable juices, liquid or concrete, which consist of resin combined with a peculiar essential oil called oil of turpentine.

There are several varieties of turpentine on the market namely: White, European, Canada, Venice and China, all of which resemble one another in odor and taste, but are distinguished by different shades.

Soluble in alcohol and unites with the fixed oils.

It is stimulant diuretic, anthelmintic and laxative.

TEREBINTHINA CANADENSIS (Canada Turpentine) Balsam of Fir.

A liquid oleoresin obtained from Abies balsamea (nat. ord. Coniferæ) by puncturing or incising the bark of the trunk and branches. It grows in Canada and the State of Maine. It contains resin and volatile oil. Soluble in ether, chloroform or benzol. By time and exposure it becomes thicker and more yellow, and finally solid. Official in flexible collodion.

ZINGIBER (Ginger).

The rhizome of Zingiber officinale (nat. ord. Scitamineæ), growing in India, Africa and the West Indies. The Jamaica or white ginger is the best quality. Contains Volatile Oil and resin.

Powdered ginger is often adulterated with rice starch, brick dust and chalk. Official in the fluid extract, oleoresin, aromatic powder, tincture, and compound powder of rhubarb. It is a stimulant and carminative.

EXTRACTUM ZINGIBERIS FLUIDUM (Fluid Extract of Ginger).

Prepared in the usual manner from ginger No. 40, using alcohol as the menstruum.

Dose: 10 to 20 minims (0.6 to 1.25 C.c.). Official in the syrup.

OLEORESINA ZINGIBERIS (Oleoresin of Ginger).

Prepared by percolating ginger No. 60 with ether, recovering the ether, and keeping in well-stoppered bottles. Dose: 1 minim (0.06 C.c.).

TINCTURA ZINGIBERIS (Tincture of Ginger).

Prepared by macerating and percolating ginger No. 40 with alcohol. Dose: 8 to 40 minims (0.5 to 2.5 C.c.) Official in the troche.

TROCHISCI ZINGIBERIS (Troches of Ginger).

Contain tincture of ginger, tragacanth, sugar and syrup of ginger. Each troche contains 3 minims of the tincture.

SYRUPUS ZINGIBERIS (Syrup of Ginger).

Prepared by triturating fluid extract of ginger with precipitated calcium phosphate, allowing the alcohol to evaporate, triturating with water, filtering, and dissolving sugar in the filtrate without heat.

Dose: 1 fluid drachm (3.7 C.c.).

BALSAMS.

BENZOINUM (Benzoin) Gum Benjamin.

A balsamic resin obtained from Styrax Benzoin (nat. ord. Styraceæ), usually by making deep incisions in the bark of the trees, and allowing the liquid that exudes to concrete on exposure to the air. It grows in Sumatra, Java, Borneo, and Siam.

Contains volatile oil, resin, benzoic and cinnamic acids.

Official in benzoinated lard, tincture and compound tincture. It is a stimulant and expectorant.

TINCTURA BENZOINI (Tincture of Benzoin).

Prepared by macerating benzoin in moderately coarse powder, with alcohol for seven days and filtering.

Dose: 20 to 30 minims (1.25 to 1.9 C.c.).

TINCTURA BENZOINI COMPOSITA (Compound Tincture of Benzoin) Elixir Traumaticum, Commander's Balsam, Wade's Balsam, Friar's Balsam, Jesuit Drops, Turlington's Balsam, Balsam of Life.

Contains benzoin in coarse powder, purified aloes, storax, balsam of tolu and alcohol.

It is a stimulant expectorant. Dose: 30 to 120 minims (1.9 to 7.5 C.c.).

BALSAMUM PERUVIANUM (Balsam of Peru).

A balsam obtained from Toluifera Pereiræ (nat. ord. Leguminosæ) exuding from the trunks of the trees after the bark has been beaten, scorched and removed. It grows in Central America.

Contains resin, volatile oil, benzoic and cinnamic acids.

On exposure to the air it does not become hard. It mixes with absolute alcohol, chloroform, glaciale acetic acid. Partially soluble in ether and benzin. Often adulterated with castor oil, copaiba, Canada Turpentine, etc.

It is a warm stimulating stomachic. Dose: ½ to 1 fluid drachm (1.9 to 3.75 C.c.).

BALSAMUM TOLUTANUM (Balsam of Tolu).

A balsam obtained from Toluifera Balsamum (nat. ord. Leguminoseæ), exuding from the trunks of the trees. It grows in Carthagena.

Contains volatile oil, resin and free acid.

By age it becomes hard and brittle like resin. Soluble in oils and alcohol, chloroform, solutions of fixed alkalies, ether; nearly insoluble in water, benzin or carbon disulphide.

Official in the tincture, syrup, and compound tincture of benzoin.

TINCTURA TOLUTANI (Tincture of Tolu).

Prepared by dissolving balsam of tolu in alcohol and filtering.

Dose: 1 to 2 fluid drachms (3.75 to 7.5 C.c.).

SYRUPUS TOLUTANUS (Syrup of Tolu).

Prepared by dissolving balsam of tolu in alcohol, mixing with precipitated calcium phosphate and sugar, setting aside in a warm place until the alcohol has evaporated, adding water, filtering, and dissolving more sugar in the filtrate.

It is used chiefly to impart an agreeable flavor to preparations.

STYRAX (Storax).

A balsam obtained from the inner bark of Liquidamber orientalis (nat. ord. Hamamelaceæ) purified by solution in spirit, filtering and straining. It grows in Asia Minor. Contains volatile oil, resin, benzoic and cinnamic acids.

Often adulterated with turpentine. Soluble in alcohol and ether, but insoluble in water.

Official in compound tincture of benzoin. It is a stimulating expectoraut. Dose: 10 to 20 grains (0.65 to 1.3 gm.).

Astringent Drugs.

CASTANEA (Chestnut).

The leaves of Castanea dentata (nat. ord. Cupuliferæ) growing in the United States and Europe.

Contains tannin, chlorophyll, gallic acid, gum, albumen.

Official in the fluid extract. Used in the treatment of whooping cough.

EXTRACTUM CASTANEÆ FLUIDUM (Fluid Extract of Castaneæ).

Prepared by percolating chestnut leaves No. 30 with boiling water until exhausted, evaporating to a small bulk, filtering, and adding alcohol and glycerin. Dose: 1 to 2 fluid drachms (3.75 to 7.5 C.c.).

CATECHU (Catechu) Cutch, Terra Japonica.

An extract prepared from the wood of Acacia Catechu (nat. ord. Leguminosæ), growing in the East Indies and Hindoostan.

Contains from 45% to 55% tannic acid, catechin, etc. It is often mixed with sand, sticks and other impurities. The variety of tannic acid called catechu-tannic acid is not, like the tannic acid of galls, converted into gallic acid by exposure to the air.

Official in the compound tincture and troche.

It is a powerful astringent. Dose: 10 to 30 grains (0.65 to 1.95 gm.).

TINCTURA CATECHU COMPOSITA (Compound Tincture of Catechu).

Prepared by percolating without macerating, a mixture of catechu No. 40, cassia cinnamon No. 40, with dilute alcohol.

Dose: 30 minims to 3 fluid drachms (1.9 to 11.25 C.c.).

TROCHISCI CATECHU (Troches of Catechu).

Contains catechu in fine powder, tragacanth and stronger orange flower water.

Each troche contains about 1 grain (0.065 gm.) of catechu.

GALLA (Nutgall).

An excrescence on Quercus Lusitanica (nat. ord. Cupuliferæ), caused by the punctures and deposited ova of Cynips Gallæ tinctoriæ. It is obtained from Syria and Asia Minor.

Contains from 65% to 77% of tannic acid (called gallo-tannic acid).

Official in the tincture and ointment. It is an astringent, but is not used internally.

TINCTURA GALLÆ (Tincture of Gall).

Prepared by percolating without macerating, gall No. 40, with a mixture of alcohol and glycerin.

Dose: 1 to 3 fluid drachms (3.7 to 11.25 C.c.).

UNGUENTUM GALLÆ (Ointment of Gall).

Prepared by rubbing gall No. 80 with benzoinated lard.

Used chiefly for piles, etc.

GERANIUM (Cranesbill).

The rhizome of Geranium maculatum (nat. ord. Geraniaceæ), growing in the United States.

Contains tannic acid, gallic acid, mucilage, resin, etc.

Official in the fluid extract. It is an astringent.

EXTRACTUM GERANII FLUIDUM (Fluid Extract of Geranium).

Prepared in the usual manner from geranium No. 30, using a mixture of dilute alcohol and glycerin as the menstruum.

Dose: 30 to 60 minims (1.9 to 3.7 C.c.).

HÆMATOXYLON (Logwood).

The heart wood of Hæmatoxylon Campechianum (nat. ord. Leguminosæ), growing in Campechy and Jamaica.

Contains a coloring principle Hæmatoxylin, tannin, resinous matter, etc. Official in the extract. It is a mild astringent.

EXTRACTUM HÆMATOXYLI (Extract of Hæmatoxylon).

Prepared by macerating logwood (rasped) with water, boiling and evaporating to dryness. Dose: 10 to 30 grains (0.65 to 0.95 gm.).

HAMAMELIS (Witch Hazel).

The leaves of Hamamelis Virginiana (nat. ord. Hamamelaceæ), collected in autumn, and growing all over the United States.

Contains tannin, resin, extractive, volatile oil, etc.

Official in the fluid extract. It is a mild astringent.

EXTRACTUM HAMAMELIDIS FLUIDUM (Fluid Extract of Hamamelis).

Prepared in the usual manner from hamamelis No. 40, using a mixture of alcohol, glycerin and water as the menstruum.

Dose: ½ fluid drachm (1.9 C.c.).

KINO (Kino).

The inspissated juice of Pterocarpus Marsupium (nat. ord. Leguminosæ), growing in the East Indies, West Indies, South America, Africa, Australia; that from the East Indies being the official variety.

Contains tannic acid, gallic acid, pectin, etc. Official in the tincture. It is an astringent.

TINCTURA KINO (Tincture of Kino).

Prepared by macerating and filtering kino with a mixture of alcohol, glycerin, and water. The glycerin is used to prevent gelatinization.

Dose: 1 to 2 fluid drachms (3.75 to 7.5 C.c.).

KRAMERIA (Rhatany).

The root of Krameria triandra, and of Krameria Ixina (nat. ord. Polygaleæ) growing in Peru.

Contains tannic acid (called krameria-tannic acid), gum, starch, etc.

Official in the extract, fluid extract and tincture.

It is a tonic and astringent. Dose. 20 to 30 grains 1.3 to 1.95).

EXTRACTUM KRAMERIÆ (Extract of Krameria).

Prepared by boiling krameria No. 40 with water, and evaporating to dryness. Official in the troche. Dose: 10 to 20 grains (0.65 to 1.3 gm.).

TROCHISCI KRAMERIÆ (Troches of Krameria).

Contain extract of krameria, sugar, tragacanth, and orange flower water.

EXTRACTUM KRAMERIÆ FLUIDUM (Fluid Extract of Krameria).

Prepared in the usual manner from krameria No. 30, using a mixture of dilute alcohol and glycerin as the menstruum.

Official in the syrup. Dose: 10 to 60 minims (0.6 to 3.75 C.c.).

SYRUPUS KRAMERIÆ (Syrup of Krameria).

Prepared by mixing together fluid extract of krameria and syrup.

Dose for adults: ½ fluid ounce (15 C.c.). For children: 20 to 30 drops (1.25 to 1.9 C.c.).

LIMONIS SUCCUS (Lemon Juice).

The freely expressed juice of the ripe fruit of Citrus Limonum (nat. ord. Rutaceæ). Contains citric acid, upon which its properties depend.

It is refrigerant. Used by sea-faring men to prevent scurvy. Dose: 1 to 4 fluid ounces (30 to 120 C.c.).

QUERCUS ALBA (White Oak).

The bark of Quercus Alba (nat. ord. Cupuliferæ), growing in the United States. Contains tannic acid, gallic acid, and extractive matter.

It is astringent and tonic, but is not used internally.

RHUS GLABRA (Sumach).

The fruit of Rhus glabra (nat. ord. Anacardieæ), growing in the United States.

Contains tannic acid, gallic acid, albumen, gum, starch, caoutchouc, resin, etc. Official in the fluid extract. It is astringent and refrigerant.

EXTRACTUM RHOIS GLABRÆ FLUIDUM (Fluid Extract of Rhus Glabra).

Prepared in the usual manner from rhus glabra No. 40, using a mixture of dilute alcohol and glycerin as the menstruum.

Used principally as an addition to mouth washes and gargles.

ROSA GALLICA (Red Rose) Red Rose Petals.

The petals of Rosa Gallica (nat. ord. Rosaceæ), collected before expanding. It grows in the United States and Europe.

Contains tannin, gallic acid, coloring matter, volatile oil, fixed oil, albumen, etc.

Official in the confection, fluid extract and pill of aloes and mastic.

CONFECTIO ROSÆ (Confection of Rose).

Contains red rose No. 60, sugar, clarified honey, stronger orange flower water.

It is slightly astringent, but is used mostly as a vehicle for other medicines.

EXTRACTUM ROSÆ FLUIDUM (Fluid Extract of Rose).

Prepared in the usual manner from red rose No. 30, using a mixture of dilute alcohol and glycerin as the menstruum.

Official in honey of rose and the syrup.

Dose: 1 to 2 fluid drachms (3.7 to 7.5 C.c.).

MEL ROSÆ (Honey of Rose).

Contains fluid extract of rose and clarified honey.

Used as an addition to gargles.

SYRUPUS ROSÆ (Syrup of Rose).

Contains fluid extract of rose and syrup.

Dose: 1 fluid drachm (3.7 C.c.).

RUBUS (Blackberry).

The bark of the root of Rubus villosus, Rubus Canadensis and Rubus trivialis (nat. ord. Rosaceæ), growing in the United States.

Contains tannic acid upon which its properties depend.

Official in the fluid extract. It is tonic and astringent.

EXTRACTUM RUBI FLUIDUM (Fluid Extract of Rubus).

Prepared in the usual manner from rubus No. 60, using a mixture of alcohol, water and glycerin as the menstruum.

Official in the syrup. Dose: $\frac{1}{2}$ to 1 fluid drachm (1.9 to 3.7 C.c.).

SYRUPUS RUBI (Syrup of Rubus).

Prepared by mixing fluid extract of rubus with syrup.

Dose: 1 to 2 fluid drachms (1.9 to 3.7 C.c.).

TAMARINDUS (Tamarind).

The preserved pulp of the fruit of tamarindus Indica (nat. ord. Leguminoseæ), growing in the East and West Indies.

Contains citric acid, tartaric acid, malic acid, potassium bi-tartrate, gum, etc. Official in confection of senna. It is a refrigerant and laxative.

VIBURNUM OPULUS (Cramp Bark).

The bark of Viburnum Opulus (nat. ord. Caprifoliaceæ), growing in the United States. Contains tannic acid.

Official in the fluid extract. Little used in medicine.

EXTRACTUM VIBURNI OPULI FLUIDUM (Fluid Extract of Cramp Bark).

Prepared in the usual manner from cramp bark No. 60, using a mixture of 750 parts alcohol and 250 parts water as the menstruum.

Dose: $\frac{1}{2}$ to 1 fluid drachm (1.9 to 3.7 C.c.).

ANIMAL PRODUCTS.

Acidum Lacticum. See page 91.

Adeps. See page 114.

OLEUM ADIPIS (Lard Oil).

A fixed oil expressed from lard at a low temperature. At a temperature a little below 10°C (50°F) it usually commences to deposit a white granular fat, and at near 0°C (32°F) it forms a semi-solid, white mass. It consists principally of olein with a small quantity of stearin, and is sometimes adulterated with paraffine oil.

Used for the purpose of making ointment of mercuric nitrate.

Adeps Lanæ Hydrosus. See page 114.

CANTHARIS (Spanish Flies).

The dried beetle of Cantharis vesicatoria (class, Insecta: order, Coleoptera) obtained from Spain, Italy and France, and should be thoroughly dried at a temperature not exceeding 40°C (104°F) and kept in well-closed bottles.

Contains a crystalline substance Cantharidin, a green oil, a black matter, a yellow viscid matter, fatty matter, etc.

If the powdered flies be kept in well-stoppered bottles, they will keep their activity for years, but if exposed to damp air they will decompose.

Official in the cerate, tincture and cantharidal collodion.

Dose: 1 to 2 grains (0.065 to 0.13 gm.).

LECTURE No. 26.

CERATUM CANTHARIDIS (Cerate of Cantharides) Cantharides Plaster, Blistering Plaster.

Contains cantharides No. 60, yellow wax, resin and lard.

Official in pitch plaster with cantharides.

TINCTURA CANTHARIDIS (Tincture of Cantharides).

Prepared by macerating and percolating cantharides No. 60 with alcohol. Dose: 3 to 10 drops (0.09 to 0.3 C.c.).

Cera Alba. See page 106.

Cera Flava. See page 106.

Cetaceum. See page 106.

COCCUS (Cochineal).

The dried female of Coccus cacti (class, Insecta: order, Hemiptera), obtained from Mexico, Central America, West Indies and the southern part of the United States.

Contains a glucoside Carminic Acid $C_{17}H_{18}O_{10}$, which is decomposed by dilute sulphate acid into a non-crystallizable sugar and carmine red $C_{11}H_{12}O_7$. Often adulterated with heavy substances as talc, lead carbonate, and barium sulphate. Used as a coloring.

FEL BOVIS (Oxgall).

The fresh bile of Bos Taurus (class, Mammalia: order, Ruminantia).

It is a brownish green, somewhat viscid liquid, having a peculiar, unpleasant odor, and a disagreeable bitter taste. Official in purified oxgall.

FEL BOVIS PURIFICATUM (Purified Oxgall).

Prepared by evaporating oxgall, adding alcohol, macerating, filtering to remove mucilaginous matter and evaporating to a pilular consistence.

Soluble in water and alcohol. It is tonic and laxative.

Dose: 5 to 10 grains (0.33 to 0.65 gm.).

ICHTHYOCOLLA (Isinglass) Fish Glue.

The swimming bladder of Acipenser Huso, and of other species of Acipenser (class, Pisces: order, Sturiones), obtained principally from Russia.

It is the purest form of gelatin and swells up when added to cold water, but boiling water dissolves it. Insoluble in alcohol

Official in the plaster.

EMPLASTRUM ICHTHYOCOLLÆ (Court Plaster).

Prepared by dissolving isinglass in hot water, spreading one-half of it on taffeta, mixing the second half with alcohol and glycerin, applying the same as the first, reversing the taffeta, and coating with tincture of benzoin, and drying.

MOSCHUS (Musk).

The dried secretion from the preputial follicles of Moschus moschiferus (class, Mammalia: order, Ruminantia), obtained from Central Asia.

About 10% of it is soluble in alcohol and 50% is soluble in water. It is inflammable.

Often adulterated with dried blood, sand, lead, iron filings, hair, animal membrane, tobacco, dung of birds, wax, benzoin, asphaltum, and artificial musk. Official in the tincture. It is a stimulant and antispasmodic.

TINCTURA MOSCHI (Tincture of Musk).

Prepared by rubbing musk with water, adding alcohol, macerating 7 days, filtering and making up to the required quantity with dilute alcohol.

Dose: 30 to 120 minims (1.9 to 7.5 C.c.).

Oleum Morrhuæ. See page 115.

Pancreatinum. See page 93.

Pepsinum. See page 93.

Saccharum Lactis. See page 110.

Sevum. See page 118.

VITELLUS (Yolk of Egg).

The yolk of the egg of Gallus Bankiva (class, Aves : order, Gallinæ). Official in the glycerite.

GLYCERITUM VITELLI (Glycerite of Yolk of Egg), Glyconin.

Prepared by rubbing together glycerin and yolk of egg.

Used as an emulsifying agent and as an application to burns, etc.

TABLE OF SOME OF THE MORE COMMON INCOMPATIBLES.

ACACIA.—Alcohol, soluble lead salts, tincture chloride of iron, concentrated solution of borax.

ACETANILID.—Ferric Salts.

ACID, ARSENOUS.—Tannic acid, lime water, salts of iron and magnesium, chromic acid, silver and gold salts.

ACID, BENZOIC.—Silver nitrate, corrosive sublimate, ferric salts, chlorine, lead acetate.

ACID, CARBOLIC.—Albumen, chloral, collodion, salts of iron, chlorine, bromine, nitric acid.

ACID, CITRIC.—Alkaline solutions, whether pure or carbonated (converting them into citrates), earthy and metallic carbonates, most acetates, alkaline sulphides and soaps, sodium salicylate.

ACID, GALLIC.—Lime water, silver nitrate, tartar emetic, ferrous sulphate, ferrous iodide, lead acetate, potassium carbonate.

ACID, HYDROCHLORIC.—Alkalies, metallic oxides, sulphides, potassium tartrate, tartar emetic, silver nitrate, lead compounds, etc.

ACID, HYDRIODIC, SYRUP.—Potassium chlorate, mineral acids, salts of the metals.

ACID, NITRIC.—Organic matter.

ACID, NITROHYDROCHLORIC, DIL.—Strong alcoholic liquids.

ACID, PHOSPHORIC.—Ferric Pyrophosphate, silver nitrate.

ACID, SALICYLIC.—Acids, salts of the metals, and many alkaloids (in aqueous solution), lime water, hydrochloric acid.

ACID, TANNIC.—Salts of the metals, alkaloidal solutions, potassium chlorate, solutions of starch, albumen, gelatin, gluten, mineral acids, antipyrine.

ALKALOIDS.—Tannin and substances containing it, mercuric chloride, iodine and its compounds, alkalies, alkaline earths, Donovan's solution, picric acid, gold salts, bromides, acetic and acetates, salicylates.

ALUM.—Alkalies and their carbonates, lime and lime water, magnesia and its carbonate, potassium tartrate, lead acetate, iron and zinc sulphates.

AMMONIA.—Acids, alum, mineral salts, alkaloids.

AMMONIUM ACETATE.—Alkalies, strong acids, mercuric chloride, silver nitrate.

AMMONIUM BROMIDE.—Acids, acid salts, spirit of nitrous ether.

AMMONIUM CARBONATE.—Acids, potassa, magnesia, alum, calcium chloride, potassium, bitartrate, mercuric chloride, salts of iron and lead, zinc sulphate.

ANTIMONY AND POTASSIUM TARTRATE.—Acids, alkalies and their carbonates, some of the alkaline earths and metals, calcium chloride, lead acetate, lead subacetate astringent infusions and decoctions, alcohol. mercurous chloride, soap and acids.

BISMUTH SUBNITRATE.—Ferrous sulphate, tannin, chlorinated lime, mercurous and mercury salts, sulphides, potassium iodide alkaline carbonates.

BROMINE.—Chlorine, vegetable colors, albumen, starch paste.

CALCIUM HYPOPHOSPHITE.—Soluble salts of mercury, copper and silver.

CALCIUM HYDRATE (Lime Water).—Carbonates, acids, salts of the alkalies, ammonium salts, and mercuric salts.

CALX CHLORATA.—Mineral acids, carbonic acid, alkaline carbonates.

CAMPHOR.—Chloral, bromine, sulphuric acid, nitric acid, resins.

CHLORAL —Carbolic acid, camphor, alkaline carbonates, potassium cyanide, mercurous and mercuric salts, potassium chlorate, nitric acid, sulphuric acid, alkaline salts.

FERRIC CHLORIDE, TINCTURE.—Phosphoric acid and alkaline phosphates, ferrocyanides, creosote, alkalies, alkaline earths or carbonates (with production of the hydrate or oxide), tannic acid (inky mixtures), mercurous salts (mercuric compounds), mucilage of acacia (forming a jelly), antipyrine, salicylates.

FERROUS SULPHATE.—Lime water, calcium and barium chloride, alkalies and their carbonates, soap, silver nitrate, borax, sodium phosphate, lead acetate and subacetate.

GLYCERIN.—Chlorine, and chlorinated compounds, potassium permanganate, chromic acid, borax, silver nitrate, nitric acid.

HYDROGEN DIOXIDE, SOLUTION.—Arsenous acid, ferrous salts, alkalies.

IODINE.—Starch, metallic salts, ammonia water, chloral, alkaline carbonates, acacia, sulphides and hyposulphites, alkaloids.

IODOFORM.—Silver nitrate, mercurous chloride, alkaline hydrates.

LEAD ACETATE.—Alkalies and their soluble salts the acids of which produce with lead, insoluble or sparingly soluble compounds (as citric, sulphuric, hydrochloric, and tartaric acids), lime water, potassium iodide, tannic acid, soap.

LEAD SUBACETATE, SOLUTION.—Solutions of gum, tannin, most vegetable coloring matter, many animal substances particularly albumen.

MAGNESIUM CARBONATE.—Acids, potassa, soda, lime, barium oxide, strontium oxide, acidulous and metallic salts.

MAGNESIUM SULPHATE.—Sodium phosphate, potassa, soda, and their cabonates, lime, barium and strontium oxides, and their soluble salts, ammonia, lead acetate, calcium chloride.

MERCURIC CHLORIDE.—Alkalies, sulphurous acid, hypophosphorus acid or their salts, lime water, alkaloids, silver nitrate, iodides, tannic acid, carbonates, tartar emetic, salts of iron, copper and lead, vegetable astringents.

MERCUROUS CHLORIDE.—Vegetable extracts, silver nitrate, acids, acid salts, lime water, alkaline carbonates, iodine and soluble iodides, ammonium chloride, sodium chloride, potassium chloride, hydrochloric acid, tincture of ferric chloride, or other soluble chlorides (forming mercuric chloride), antipyrine, bromides.

MERCUROUS IODIDE.—Potassium iodide, sodium chloride, mineral acids.

MERCURIC IODIDE.—Potassium iodide, sodium chloride.

METHYL SALICYLATE.—Alkalies.

PANCREATIN.—Acids.

PEPSIN.—Tannic acid, alcohol, alkalies.

PHOSPHORUS.—Potassium chlorate, chlorine, bromine, iodine, carbon disulphide, sulphur.

POTASSA, SOLUTION.—Solutions of metals, alkaloids, acid solutions.

POTASSIUM ACETATE.—Mineral acids, sodium and magnesium acetates, mercuric chloride, silver nitrate.

POTASSIUM ARSENITE, SOLUTION.—Silver nitrate, ferrous iodide, ferrous sulphate, calcium chloride, acids, lime water, alum, vegetable astringents, alkaloids.

POTASSIUM BROMIDE.—Solutions containing free chlorine, nitrous or nitric acids, alkaloids, calomel.

POTASSIUM CARBONATE.—Acids and acidulous salts, ammonium chloride, ammonium acetate, lime water, calcium chloride, magnesium sulphate, alum, silver nitrate, ammoniated copper, lead acetate, lead subacetate, zinc sulphate.

POTASSIUM CHLORATE.—Organic matter, sulphur, antimony sulphide, phosphorus.

POTASSIUM IODIDE.—Acids acid salts, silver nitrate, tartaric acid, salts of mercury, iron and lead, alkaloids, mercurous chloride, mercurous and mercuric oxides, turpeth mineral, white precipitate, blue mass, spirit of nitrous ether, bismuth subnitrate.

POTASSIUM PERMANGANATE.—Phosphorus, hypophosphites, iodine, iodides, silver nitrate, organic matter.

POTASSIUM AND SODIUM TARTRATE.—Most acids, all acidulous salts, except potassium bitartrate, lead acetate, lead subacetate, soluble salts of barium and calcium.

SILVER NITRATE.—Albumen, acetic acid, hydrocyanic acid, iodides, bromides, potassium ferrocyanide, most all spring and river water (on account of a little salt that they may contain), soluble chlorides, sulphuric acid, hydrogen sulphide, hydrochloric acid, tartaric acid, and their salts, alkalies and their carbonates, lime water, astringent infusions.

SODIUM CARBONATE.—Acids, acidulous salts, lime water, ammonium chloride, earthy and metallic salts.

SODIUM CHLORIDE.—Some of the acids (particularly sulphuric and nitric), silver nitrate, mercurous nitrate.

SODA CHLORINATED, SOLUTION.—Metallic salts, iodides, bromides.

SODIUM PHOSPHATE.—Soluble salts of lime, neutral metallic solutions, alkaloids.

SODIUM SALICYLATE.—Acids, salts of the metals, solutions of many alkalies, antipyrine, alkaloids.

SODIUM SULPHATE.—Potassium carbonate, calcium chloride, salts of barium, lead acetate, lead subacetate, strong solutions of silver.

STRONTIUM BROMIDE.—Soluble carbonates and sulphates, all other bromides, mineral acids.

ZINC ACETATE.—Mineral Acids.

ZINC PHOSPHIDE.—Vegetable extracts

ZINC SULPHATE.—Alkalies, alkaline carbonates, sulphides, lime water, soluble salts of lead, astringent infusions, barium chloride, potassium ferrocyanide.

TABLE OF ABBREVIATIONS AND CONTRACTIONS FREQUENTLY USED IN PRESCRIPTION WRITING.

Aa. (*ana*) of each ingredient.

Abs. febr. (*absente febre*) in the absence of the fever.

Ad. libit. (*ad libitum*) at pleasure.

Add. (*adde or addantur*) add or let be added.

Adst. febre (*adstante febre*) when the fever is on.
Aggred. febre (*aggrediente febre*) while the fever is coming on.
Altern. horis (*alternis horis*) every other hour.
Appl. (*applicetum*) let them be applied.
Aq. bull. (*aqua bulliens*) boiling water.
Aq. com. (*aqua communis*) common water.
Aq. ferv. (*aqua fervens*) hot water.
Aq. font. (*aqua fontana*) spring water.
Bis ind. (*bis indies*) twice a day.
C. (*cum*) with.
Coch. (*cochleare*) a spoonful.
Coch. ampl. (*cochleare amplum*) a large spoonful.
Coch. infant. (*cochleare infantis*) a child's spoonful.
Chart. (*charta*) paper.
Colet. (*coletur*) let it be strained.
Comp. (*compositus*) compound.
Cong. (*congius*) a gallon.
Coq. (*coquo*) boil.
Dec. (*decanta*) pour off.
Det. (*detur*) let it be given.
Dieb. alt. (*diebus alternis*) every other day.
Dieb. tert. (*diebus tertiis*) every third day.
Dil. (*dilutus*) dilute or diluted.
Dim. (*dimidius*) one-half.
D. (*dosis*) dose.
Ejusd. (*ejusdem*) of the same.
F. (*fac*) make.
Fiat—let it be made.
Feb. dur. (*febre durante*) during the fever.
Fl. (*fluidus*) liquid.
F. M. (*fiat mistura*) let a mixture be made.
F. S. A. (*fiat secundum artem*) let it be made according to art.
Gr. (*granum*) grains.
Gtt. (*gutta*) a drop.
H. S. (*hora somni*) just before going to sleep.
Hor. un. spatio (*horæ unius spatio*) at the expiration of an hour.
Hor. interm. (*horis intermediis*) at the intermediate hours.
Ind. (*indies*) from day to day, or daily.
Int. (*inter*) between.
M. (*misce*) mix.
Mic. Pan. (*mica panis*) crumbs of bread.
Mitt. (*mitte*) send.
Mod. præso. (*modo præscripto*) in the manner prescribed.
More dict. (*more dicto*) in the manner directed.

Mor. sol. (*more solito*) in the usual manner.
O. (*octarius*) a pint.
Omn. hor. (*omni hora*) every hour.
Omn. bid. (*omni biduo*) every two days.
Omn. bih. (*omni bihorio*) every two hours.
O. M. (*omni mane*) every morning.
O. N. (*omni nocte*) every night.
Omn. quadr. hor. (*omni quadrante horæ*) every quarter of an hour.
P. æ. (*part æqual*) equal parts.
P. C. (*post cibum*) after eating.
P. r. n. (*pro re nata*) occasionally.
Pulv. (*pulvis*) a powder.
Q. s. (*quantum sufficiat*) much as is sufficient.
Repet. (*repetatur*) let it be continued.
S. A. (*secundem artem*) according to art.
Semih. (*semihora*) half an hour.
Sesquih. (*sesquihora*) an hour and a half.
St. (*stent*) let it stand.
Stat. (*statim*) immediately.
Sum. tal. (*summat talem*) let the patient take one like this.

Rules for Converting Thermometric Scales.

1—To convert Centigrade degrees into those of Fahrenheit above 32, multiply by 1.8 and add 32. Ex. 50 degrees Centigrade into those of Fahrenheit: $50 \times 1.8 = 90 + 32 = 122$.

2—To convert Fahrenheit degrees above 32 into those of Centigrade, subtract 32 and divide by 1.8. Ex. 40 degrees Fahrenheit into those of Centigrade: $40 - 32 = 8 \div 1.8 = 4.4$ degrees.

Fineness of Powders.

A No. 80 powder is a very fine powder.
A No. 60   "   " " fine powder.
A No. 50   "   " " moderately fine powder.
A No. 40   "   " " moderately coarse powder.
A No. 20   "   " " coarse powder.

The different numbers mean a powder that has passed through a sieve having the corresponding number of meshes to the linear inch.

Alligation.

To find out the quantities of drugs of different strengths to be used to make a mixture of definite strengths, and also of liquids where no change in volume takes place when united.

RULE.—Draw a perpendicular line; to the left of the line write the required per cent. (or the per cent. sought); to the right of the line write the given percentages under each other. Connect each per cent. that is greater than the one sought with one that is smaller, by a line, and each one that is smaller with one that is greater than the per cent. sought.

Opposite each given per cent. write the difference between the given per cent. at the other end of the line with which it is connected, and the per cent. sought. The corresponding figures will give the number of parts by weight of each ingredient to use. Ex. We have a lot of powdered opium containing 12% of morphine, and one containing 15% : in what proportion must they be united to obtain a mixture containing 14%?

```
                                        Proof.
      | 12 ⎤          1 part of 12%     1×12=12
  14  |
      | 15 ⎦   2      2 parts of 15%    2×15=30
                      3 parts.          3      3|42
                                                14
```

We have four lots of powdered opium, containing respectfully 10%, 12%, 15%, and 16% of morphine; how much of each must be used to make a mixture containing 14%?

```
                                        Proof.
      | 10 ⎤   2      2 parts of 10%    2×10=20
      | 12 |   1      1 part  of 12%    1×12=12
  14  | 15 |   2      2 parts of 15%    2×15=30
      | 16 ⎦   4      4 parts of 16%    4×16=64
                      9                 9      9|126
                                                14
```

The given percentages may be connected in various ways as long as one smaller than the required percentage is connected with one larger, but each per cent. must be connected with some other.

Example as above.

```
                                        Proof.
      | 10 ⎤   1      1 part  of 10%    1×10=10
      | 12 |   2      2 parts of 12%    2×12=24
  14  | 15 |   4      4 parts of 15%    4×15=60
      | 16 ⎦   2      2 parts of 16%    2×16=32
                      9 parts           9      9|126
                                                14
```

If there should be an uneven number of given per cents., either above or below the required per cent., any of them may have more than one line connected with it if necessary, and the corresponding numbers which would be obtained should be added together.

Ex. We have four lots of powdered opium containing respectfully 10%, 12%, 13% and 16% of morphine; how much of each should be used to obtain a mixture containing 14%?

```
                                             Proof.
      | 10 ⎤   2         2 parts of 10%      2×10= 20
      | 12 |   2         2 parts of 12%      2×12= 24
  14  | 13 |   2         2 parts of 13%      2×13= 26
      | 16 ⎦   1+2+4=7   7 parts of 16%      7×16=112
                         13 parts.           13     13|182
                                                     14
```

APPENDIX.

Equivalents of the Metric System in our Systems of Weights and Measures.

Rules for converting one system into that of another.

To convert Metric Weights or Measures into those of our system.

Rule: Multiply the Metric quantities by the corresponding equivalent.

Meters into inches × 39.370
Liters into fluid ounces × 33.815
Cubic centimeters into minims × 16.230
Grams into grains × 15.432
Grams into Avoir. ounces × ... 28.35
Grams into Troy ounces × 31.103
Cubic centimeters into fluid ounces × 0.0338

To convert our system of weights and measures into those of Metric.

Rule: Divide the quantities by the corresponding Metric equivalent.

Inches into Meters ÷ 39.370
Fluid ounces into Liters ÷ .. 33.815
Minims into Cubic centimeters ÷ 16.230
Grains into Grams ÷ 15.432
Avoir. ounces into grams ÷ .. 28.35
Troy ounces into grams 31.103
Fluid ounces into Cubic centimeters ÷ 0.0338

Value of Apothecaries in Metric Weight.

Grains.	Grams.	Grains.	Gms.
1-100	= 0.000648	18 =	1.166
1-64	" 0.00101	19 "	1.231
1-60	" 0.00110	20 "	1.296
1-50	" 0.00130	21 "	1.360
1-48	" 0.00135	22 "	1.425
1-40	" 0.00162	23 "	1.490
1-36	" 0.00180	24 "	1.555
1-32	" 0.00202	25 "	1.620
1-30	" 0.00220	26 "	1.685
1-25	" 0.00270	27 "	1.749
1-24	" 0.00274	28 "	1.814
1-20	" 0.00324	29 "	1.879
1-18	" 0.00360	30 "	1.944
1-16	" 0.00405	31 "	2.008
1-15	" 0.00432	32 "	2.073

Grains.	Grams.	Grains.	Gms.
1-12	= 0.00540	33 =	2.138
1-10	" 0.00648	34 "	2.203
1-8	" 0.00810	35 "	2.268
1-6	" 0.01080	36 "	2.332
1-5	" 0.01296	37 "	2.397
1-4	" 0.01620	38 "	2.462
1-3	" 0.02160	39 "	2.527
1-2	" 0.03240	40 "	2.592
3-4	" 0.04860	41 "	2.656
1	" 0.0648	42 "	2.721
2	" 0.1296	43 "	2.786
3	" 0.1944	44 "	2.851
4	" 0.2592	45 "	2.916
5	" 0.3239	46 "	2.980
6	" 0.3887	47 "	3.045
7	" 0.4536	48 "	3.110
8	" 0.5184	49 "	3.175
9	" 0.5832	50 "	3.240
10	" 0.6480	51 "	3.304
11	" 0.7129	52 "	3.368
12	" 0.7776	53 "	3.433
13	" 0.8424	54 "	3.498
14	" 0.9072	55 "	3.563
15	" 0.9719	56 "	3.638
15.432	" 1.000	57 "	3.702
16	" 1.037	58 "	3.777
17	" 1.101	59 "	3.832

Drachms.	Grams.	Drachms.	Gms.
1	= 3.888	5 =	19.440
2	" 7.776	6 "	23.328
3	" 11.664	7 "	27.216
4	" 15.552		

Ounces.	Grams.	Ounces.	Grams.
1	= 31.103	11 =	342.138
2	" 62.207	12 "	373.250
3	" 93.310	13 "	404.345
4	" 124.414	14 "	435.449
5	" 155.517	15 "	466.552
6	" 186.621	16 "	497.656
7	" 217.724	17 "	528.759
8	" 248.823	18 "	559.863
9	" 279.931	19 "	590.966
10	" 311.035	20 "	622.070

In these tables the following abbreviations will be found: oz. for ounce; grm. for gram; grn. for grain; lb. for pound; fl. dr. for fluid drachm; c. c. for cubic centimenter; fl. oz. for fluid ounce, and min. for minim.

Value of Metric in Apothecaries Weight.

Grms.		Grns.	Grms.		Grns.
0.0010	=	1-64	0.900	=	13.889
0.0013	"	1-52	1.	"	15.433
0.0015	"	1-44	2.	"	30.865
0.0020	"	1-32	3.	"	46.297
0.0025	"	1-26	4.	"	61.729
0.0030	"	1-22	5.	"	77.162
0.0035	"	1-18	6.	"	92.594
0.0040	"	1-16	7.	"	108.026
0.0045	"	1-15	8.	"	123.459
0.0050	"	1-13	9.	"	138.891
0.0055	"	1-12	10.	"	154.323
0.0060	"	1-11	11.	"	169.756
0.0065	"	1-10	12.	"	185.188
0.0070	"	1-9	13.	"	200.621
0.0080	"	1-8	14.	"	216.053
0.0090	"	1-7	15.	"	231.485
0.0108	"	1-6	16.	"	246.918
0.0162	"	1-4	17.	"	262.350
0.0324	"	1-2	18.	"	277.782
0.0486	"	3-4	19.	"	293.214
0.0567	"	7-8	20.	"	308.647
0.0648	"	1 000	21.	"	324.079
0.065	"	1.003	22.	"	339.512
0.070	"	1.080	23.	"	354.944
0.075	"	1.157	24.	"	370.376
0.080	"	1.235	25.	"	385.809
0.085	"	1.312	26.	"	401.241
0.090	"	1.389	27.	"	416.673
0.095	"	1.466	28.	"	432.106
0.100	"	1.543	29.	"	447.538
0.110	"	1.698	30.	"	462.970
0.120	"	1.852	31.	"	478.403
0.130	"	2.006	32.	"	493.835
0.140	"	2.161	33.	"	509.268
0.150	"	2.315	34.	"	524.700
0.160	"	2.469	35.	"	540.132
0.170	"	2.623	36.	"	555.565
0.180	"	2.778	37.	"	570.997
0.190	"	2.932	38.	"	586.429
0.200	"	3 086	39.	"	601.862
0.210	"	3.241	40.	"	617.294

Grms.		Grns.	Grms.		Grns.
0.220	=	3.395	50.	=	771.617
0.230	"	3.549	60.	"	925.941
0.240	"	3.704	70.	"	1080.264
0.250	"	3.858	80.	"	1234.588
0.260	"	4.012	90.	"	1388.911
0.270	"	4.167	100.	"	1543.235
0.280	"	4.321	125.	"	1929.044
0.290	"	4.475	150.	"	2314.852
0.300	"	4.630	200.	"	3086.470
0.310	"	4.784	250.	"	3858.087
0.320	"	4.938	300.	"	4629.705
0.330	"	5.093	333.	"	5144.118
0.340	"	5.247	350.	"	5401.322
0.350	"	5.401	400.	"	6172.940
0.360	"	5.556	450.	"	6944.557
0.370	"	5.710	500.	"	7716.174
0.380	"	5.864	600.	"	9259.409
0.390	"	6.019	700.	"	10802.644
0.400	"	6.173	750.	"	11574.262
0.500	"	7.716	800.	"	12345.879
0.600	"	9.259	900.	"	13889.114
0.700	"	10.803	1000.	"	15432.350
0.800	"	12.346			

Value of Avoirdupois in Metric Weight.

ozs.		grms.	ozs.		grms.
1-16	=	1.772	7	=	198.45
1-8	"	3.554	8	"	226.80
1-4	"	7.088	9	"	255.15
1-2	"	14.175	10	"	283.50
1	"	28.350	11	"	311.84
2	"	56.700	12	"	340.19
3	"	85.049	13	"	368.54
4	"	113.398	14	"	396.89
5	"	141.75	15	"	425.24
6	"	170.07			

lbs.		grms.	lbs.		grms.
1	"	453.59	6	"	2721.55
2	"	907.18	7	"	3175.14
2.2	"	1000.00	8	"	3628.74
3	"	1360.78	9	"	4082.33
4	"	1814.37	10	"	4535.92
5	"	2267.96			

Value of Metric in Avoirdupois Weight.

grms.		ozs.	grns.	grms.		ozs.	grns.
28.35	=	1		100.	=	3	230
29.	"	1	10	125	"	4	179

grms.	ozs.	grns.	grms.	ozs.	grns.
30.	= 1	25	150.	= 5	127
31.	" 1	41	200.	" 7	24
32.	" 1	56	250.	" 8	358
33.	" 1	72	300.	" 10	255
34.	" 1	87	350.	" 12	151
35.	" 1	103	400.	" 14	48
36.	" 1	118	450.	" 15	382
37.	" 1	133	500.	" 17	279
38.	" 1	149	550.	" 19	175
39.	" 1	164	600.	" 21	72
40.	" 1	180	650.	" 22	406
50.	" 1	334	700.	" 24	303
60.	" 2	50	750.	" 26	199
70.	" 2	205	800.	" 28	96
80.	" 2	360	850.	" 29	430
85.	" 3		900.	" 31	326
90.	" 3	76	950.	" 33	223
			1000.	" 35	120

Value of Apothecaries in Metric Measure.

min.	c. c.	min.	c. c.
1	= 0.06	22	= 1.36
2	" 0.12	23	" 1.42
3	" 0.18	24	" 1.48
4	" 0.25	25	" 1.54
5	" 0.31	26	" 1.60
6	" 0.37	27	" 1.66
7	" 0.43	28	" 1.73
8	" 0.49	29	" 1.79
9	" 0.55	30	" 1.85
10	" 0.62	35	" 2.16
11	" 0.68	40	" 2.46
12	" 0.74	45	" 2.77
13	" 0.80	50	" 3.08
14	" 0.86	55	" 3.30
15	" 0.92	60	" 3.70
16	" 0.00	70	" 4.31
17	" 1.05	80	" 4.93
18	" 1.11	90	" 5.54
19	" 1.17	100	" 6.16
20	" 1.23	110	" 6 78
21	" 1.29	120	" 7.30

fl. dr.	c. c.	fl. dr.	c. c.
1	= 3.70	9	= 33.27
2	" 7.39	10	" 36.97
3	" 11.09	11	" 40.66
4	" 14.79	12	" 44.36
5	" 18.48	13	" 48.06
6	" 22.18	14	" 51.75
7	" 25.88	15	" 55.45
8	" 29.57	16	" 59.10

fl. oz.	c. c.	fl. oz.	c. c.
1	= 29.57	18	= 532.26
2	" 59.14	19	" 561.93
3	" 88.67	20	" 591.50
4	" 118.34	21	" 621.08
5	" 147.81	22	" 650.65
6	" 177.39	23	" 680.22
7	" 206.96	24	" 709.80
8	" 236.53	25	" 739.37
9	" 266.10	26	" 768.94
10	" 295.68	27	" 798.51
11	" 325.25	28	" 828.09
12	" 354.82	29	" 857.66
13	" 384.40	30	" 887.23
14	" 413.97	31	" 916.80
15	" 443.54	32	" 946.38
16	" 473.11	64	" 1892.75
17	" 502.69	128	" 3785.51

Value of Metric in Apothecaries Measure.

c. c.	fl. oz.	c. c.	fl. oz.
1000	= 33.81	400	= 13.52
900	" 30.43	300	" 10.14
800	" 27.05	200	" 6.76
700	" 23.67	100	" 3.38
600	" 20.29	75	" 2.53
500	" 16.90	50	" 1.69
473	" 16.	30	" 1.01

c. c.	min.	c. c.	fl. oz.
25	= 405.7	4.	= 64.9
10	" 162.3	3.	" 48.7
9	" 146.1	2.	" 32.5
8	" 129.8	1.	" 16.23
7	" 113.6	0.50	" 8.1
6	" 97.4	0.25	" 4.1
5	" 81.1	0.06	" 1.0

The following table was arranged by E. P. Stimson, M. D., of Tiverton, R. I., for Claflin's Druggist, and is very valuable to anyone buying in the Metric weights or measures.

In making this table Dr. Stimson has somewhat extended the table of equivalents as given by Prof. Miller as compared between the pound and the Kilogram and reduced the same into its lower denominations, the figures agreeing in the main with those published by the Metric Bureau.

The first column is the price in the old weights and measures, the remaining columns the Metric values, as compared and stated in their headings, viz:

2d. column contains grains in grams, i. e., at $0.01 per grain, the gram is valued at $0.154.

3rd. column contains minims in cubic centimeters for fluids.

4th. column compares the price per ounce in avoidupois weight as stated in first column, with 100 grams.

5th. column contains the same with fluid ounce reduced to 100 c. c.

6th. and 7th. columns contain pounds or pints reduced to Kilograms or Liters.

8th. column contains fluids by the gallon to be sold by the Liter.

It would favor the use of the Metric weights if druggists would put up all their packages in Metric rather than Avoirdupois weight, and making the price in its equivalent, as per annexed tables.

This would also favor the retailer and physician, for the price paid and the price dispensed would be in the same weights; now we buy by Avoirdupois and dispense by Apothecaries or Metric, and lest we lose money we have to convert the price Avoirdupois to Apothecaries or Metric, as the case may be.

To aid in the conversion from Avoirdupois and Fluid measures to the Metric System is the object of this paper.

At. $ c	Per grains in grams.	Per minims in c.c.	Per av. oz. in 100 gms.	Per fluid oz. in 100 c. c.	Per av. lb. in kil'g'ms.	Per pint in litre.	Per gallons in litre.
.01	.154	.162	.035	.034	.022	.021	.00264
.02	.309	.324	.071	.068	.044	.042	.00528
.03	.463	.487	.106	.101	.066	.063	.00793
.04	.617	.649	.141	.135	.088	.085	.01056
.05	.772	.812	.176	.169	.110	.106	.01321
.06	.926	.974	212	.203	.132	.127	.01585
.07	1.080	1.136	.247	.237	.154	.148	.01849
.08	1.235	1.299	.282	.271	.176	.169	.02113
.09	1.389	1.461	.317	.304	.198	.190	.02378
.10	1.548	1.623	.353	.338	.220	.211	.02642
.11	1.698	1.785	.388	.372	.243	.232	.02906
.12	1.852	1.948	.423	·406	.265	.254	.03170
.13	2.006	2.110	.459	.440	.287	.275	.03434
.14	2.161	2.272	.494	.473	.309	.296	.03699
.15	2.315	2.435	.529	.507	.331	.317	.03963
16	2.469	2.597	.564	.541	.353	.338	.04227
.17	2.623	2.759	.600	.575	.375	.359	.04912
.18	2.778	2.922	.635	.609	.398	.380	.04755
.19	2.932	3.084	.670	.642	.419	.402	.05195
.20	3.086	3.246	.705	.676	.441	.423	.05284
.25	3.858	4.058	.882	.845	.551	.528	.06605

213

At. S c	Per grains in grams.	Per minims in c.c.	Per av. oz. in 100 gms.	Per fluid oz. in 100 c.c.	Per av. lb. in kil'g'ms.	Per pint in litre.	Per gallons in litre.
.30	4.630	4.869	1.058	1.014	.661	.634	.07926
.35	5.401	5.681	1.235	1.184	.772	.740	.09247
.40	6.173	6.493	1.411	1.353	.882	.845	.10567
.45	6.945	7.304	1.587	1.522	.992	.951	.11888
.50	7.716	8.116	1.764	1.691	1.102	1.057	.13209
.55	8.487	8.927	1.940	1.859	1.213	1.162	.14530
.60	9.259	9.739	2.116	2.029	1.323	1.268	.15851
.65	10.031	10.550	2.293	2.298	1.433	1.374	.17172
.70	10.803	11.362	2.469	2.367	1.543	1.479	.18493
.75	11.574	12.174	2.646	2.536	1.653	1.585	.19814
.80	12.346	12.985	2.822	2.705	1.764	1.691	.21135
.85	13.117	13.797	2.998	2.874	1.874	1.797	.22456
.90	13.889	14.608	3.175	3.043	1.984	1.902	.23777
.95	14.661	15.420	3.351	3.212	2.094	2.008	.25098
1.00	15.432	16.232	3.527	3.382	2.205	2.113	.26419
1.05	16.204	17.432	3.704	3.551	2.315	2.219	.27740
1.10	16.976	17.855	3.880	3.720	2.425	2.325	.29061
1.15	17.747	18.666	4.057	3.889	2.535	2.431	.30381
1.20	18.519	19.478	4.233	4.058	2.646	2.536	.31702
1.25	19.290	20.289	4.409	4.227	2.756	2.642	.33023
1.30	20.062	21.101	4.586	4.396	2.866	2.748	.34344
1.35	20.834	21.913	4.762	4.565	2.976	2.853	.35665
1.40	21.605	22.724	4.938	4.734	3.086	2.959	.36986
1.45	22.377	23.536	5.115	4.903	3.197	3.065	.38307
1.50	23.149	24.346	5.291	5.072	3.307	3.170	.39628
1.55	23.920	25.159	5.467	5.241	3.417	3.276	.40949
1.60	24.692	25.970	5.644	5.410	3.527	3.382	.42270
1.65	25.463	26.782	5.820	5.570	3.638	3.487	.43591
1.70	26.235	27.594	5.997	5.749	3.748	3.593	.44912
1.75	27.007	28.405	6.173	5.918	3.858	3.699	.46233
1.80	27.778	29.217	6.349	6.087	3.968	3.804	.47554
1.85	28.550	30.028	6.526	6.256	4.079	3.910	.48877
1.90	29.321	30.840	6.702	6.425	4.189	4.016	.50195
1.95	30.093	31.652	6.878	6.594	4.299	4.122	.51336
2.00	30.865	32.463	7.055	6.763	4.409	4.227	.52837
3.00	46.297	48.695	10.582	10.145	6.614	6.340	.79256
4.00	61.729	64.926	14.110	13.526	8.818	8.454	1.05674
5.00	77.162	81.158	17.637	16.908	11.023	10.567	1.32093
6.00	92.594	97.389	21.164	20.289	13.228	12.681	1.58512
7.00	108.026	113.621	24.692	23.671	15.432	14.794	1.84930
8.00	123.459	129.853	28.219	27.052	17.637	16.908	2.11349
9.00	138.891	146.084	31.747	30.434	19.842	19.021	2.37767
10.00	154.323	162.232	35.274	33.815	22.046	21.135	2.64186
20.00	308.647	324.632	70.548	67.631	44.092	42.270	5.28372
30.00	462.970	486.947	105.822	101.446	66.139	63.405	7.92558
40.00	617.294	649.263	141.096	135.261	88.185	84.540	10.56744
50.00	771.617	811.579	176.370	169.076	110.231	105.675	13.20930
60.00	925.941	973.895	211.644	202.892	132.277	126.809	15.85116
70.00	1080.264	1136.211	246.918	236.707	154.323	147.944	18.49302
80.00	1234.588	1298.526	282.192	270.522	176.370	169.079	21.13488
90.00	1388.911	1460.842	317.466	304.335	198.416	190.214	23.77674
100.00	1543.235	1623.158	352.740	338.153	220.462	211.349	26.41860
200.00	3086.470	3246.316	705.480	676.306	440.924	422.698	52.83720
300.00	4629.705	4869.474	1058.220	1014.458	661.386	634.047	105.67440
400.00	6172.939	6492.632	1410.960	1352.611	881.848	845.396	132.09300
500.00	7716.174	8115.790	1763.700	1690.764	1102.311	1056.745	158.85116
1000.00	15432.349	16231.580	3577.400	3381.528	2204.621	2113.490	264.18600

INDEX.

The Index has been condensed as much as possible. Pharmaceutical, Botanical, Therapeutical and Chemical terms will be found under their respective headings; chemical compounds are indexed under each corresponding class heading. Ex. Potassium Iodide (see potassium); the principal synonyms of drugs and chemicals are given to aid in finding them readily. The oils given in the index are those having no official drug to be classified under. All official preparations may be found under the corresponding drug or chemical, and salts of the alkaloids under the corresponding alkaloid.

	Page		Page
Acacia,	110	Aqua, Fortis,	29
Acetanilid,	104	" Regia,	28
Acetic Ether,	99	Argol,	65
Acids, Inorganic,	25	Arnica Flowers,	162
" Organic,	88	" Root,	162
Aconite,	161	Aromatic Elixir,	121
Adhesive Plaster,	189	" Powder,	125
African Pepper,	191	Arsenic,	36
Alcohol and Alcoholic Liquids	94	Arsenical Antidote,	50
Aldehyde,	101	Asafetida,	190
Aldehydes,	101	Atropine,	164
Alder Buckthorn,	147	**Balance,** Definition of	2
Alligation,	207	Balm,	131
Allspice,	133	Balsam of Life,	195
Almond, Bitter,	142	" " Peru,	195
" Sweet,	113	" " Tolu,	196
Aloe,	140	Balsams,	195
Aloin,	141	Barium Compounds,	36
Aluminum Compounds,	31	Basham's Mixture,	33
Ammonio, Ferric Alum,	48	Basilicon Ointment,	189
" Ferric Tartrate,	48	Bay Rum,	133
Ammonium Compounds,	32	Bearberry,	160
Ammoniac,	190	Bear's Weed,	186
Ammoniated Glycyrrhizin	149	Belladonna Leaf,	163
Amyl Nitrite,	100	" Root,	164
Angelica Tree,	161	Benzine Derivatives,	104
Anise,	119	Benzin,	102
Antifebrin,	104	Benzoin,	195
Antimony Compounds,	34	Beta, Napthol,	104
Apomorphine, Hydroch,	163	Bismuth Compounds,	37

	Page		Page
Bitter Almond,	142	Carron Oil,	39
Bittersweet,	171	Cascara Sagrada,	189
Blackberry,	199	Cascarilla,	143
Black Draught,	158	Cassia Fistula,	108
" Drop,	177	Cayenne Pepper,	191
" Root,	150	Celandine,	165
" Wash,	59	Cellulin,	107
Blanchard's Pills,	49	Cellulose Derivatives,	107
Blaud's Pills,	46	Cereum Compounds,	41
Bleaching Powder,	42	Chamomile, German,	131
Blistering Plaster,	201	" Roman,	120
Bloodroot,	182	Charcoal, Animal,	97
Blueberry Root,	143	" Wood,	97
Blue Butter,	58	Chemical Comb'n, Laws Regulating,	84
" Cohosh,	143	Chemical Terms, Definition of	12
" Flag,	149	Chemistry, Definition of	2
" Mass,	57	Chestnut,	196
" Ointment,	58	Chirata,	143
" Pill,	57	Chittem Bark,	189
Boneset,	147	Chlorinated Lime,	42
Borax,	73	Chlorine,	41
Botanical Terms,	19	" Derivatives,	102
Bromine,	38	Chloroform,	102
Broom Tops,	182	Chrysarobin,	144
Brown Mixture,	148	Cinnamons,	125
Bryony,	142	Cinchona,	165
Buckthorn,	147	" Red,	166
Buchu,	123	Cinchonidine,	166
Burdock,	149	Cinchonine,	166
Burgundy Pitch,	188	Citrine Ointment	58
Butterfly Weed,	142	Classes of Preparations, Defi. of	19
Cabbage Rose Petals,	133	Clove,	124
Caffeine,	172	Coca,	169
Calabar Bean,	180	Cocaine,	169
Calcined Magnesia,	55	Cochineal,	201
Calcium Compounds,	39	Codeine,	179
Calomel,	59	Cohosh, Blue,	143
Camphor,	138	" Black,	185
" Monobromated,	139	Colchicum Root,	169
Canadian Hemp,	142	" Seed,	170
Cantharides,	200	Cold Cream,	114
Caoutchouc,	186	Collodions,	107
Capsicum,	191	Colophony,	189
Caraway,	124	Columbo,	164
Carbon,	41	Commander's Balsam,	195
" Disulphide,	97	Conium,	170
Cardamom,	123	Consumptive's Weed,	186
Carragheen,	112	Copaiba,	192

	Page		Page
Copper Compounds,	42	Fig,	108
Coriander,	126	Fixed Oils,	113
Corn Silk,	161	Flake White,	54
Corrosive Sublimate,	58	Flaxseed,	115
Cotton Root Bark,	186	Fowler's Solution,	26
Couch Grass,	160	Foxglove,	146
Court Plaster,	200	Fineness of Powders,	207
Cousso,	186	Friar's Balsam,	195
Cramp Bark,	200	**Gamboge,**	191
Cranesbill,	197	Garlic,	118
Cream of Tartar,	65	Gentian,	147
Creosote,	105	German Chamomile,	131
Cubeb,	127	Ginger,	194
Cubic Nitre,	76	Glauber Salt,	77
Culver's Root,	150	Glucose,	109
" Physic,	150	Glucosides,	140
Cutch,	196	Glycerin,	96
Dandelion,	159	Glyconin,	202
Dextrine,	112	Glycerite of Boroglycerin,	27
Digitalis,	146	Glycyrrhizin, Ammon,	149
Diachylon Plaster,	55	Gold Compounds,	43
" Ointment,	55	Goldenseal,	173
Dialysed Iron,	47	Goulard's Cerate,	53
Dispensatory, Definition of	2	Gravity, Definition of	2
Donovan's Solution,	60	Green Vitriol,	50
Dover's Powder,	178	Green Soap,	115
Elaterin,	146	Gregory's Powder,	154
Elder,	135	Griffeth's Mixture,	46
Elecampane,	149	Grindelia,	172
Elements, Valence of	86	Grey Powder,	57
" Atomic Weight of	86	Guaiacum Wood,	187
Elixir, Aromatic,	121	Guarana,	172
" Pro,	141	Gums,	110
" Vitriol,	31	Gum-Arabic,	110
Elm,	113	Gum Resins,	190
English Chamomile,	120	Gun Cotton,	107
Epsom Salt,	56	**Hamamelis,**	198
Ergot,	170	Haw Black,	161
Esters,	16, 99	Hellebore, American,	184
Ether,	97	" Green,	184
" Acetic,	99	Henbane,	173
" Spirit Nitrous,	99	Hemlock,	170
Ethers,	97	Hemp, Canadian,	142
Eucalyptol,	128	" Indian,	185
Eucalyptus,	128	Honey,	109
Fats,	113	Hop,	192
Fennel,	128	Horehound,	151
Ferric Hydrate,	49	Hydrometer,	4

	Page		Page
Hydrogen Compounds,	43	Lupulin,	193
Hydrastinine,	173	Lunar Caustic,	71
Hydrous-Wool Fat,	114	**Mace,**	131
Hyoscine,	174	Magnesium Compounds,	55
Hyoscyamine,	174	Male Fern,	191
Iceland Moss,	111	Mandrake,	188
India Rubber,	186	Manganese Compounds,	56
Indian Paint,	182	Manna,	108
Indian Sage,	147	Marigold,	143
Indicators,	83	Marshmallow,	111
Iodine Compounds,	43	Mastic,	188
" Derivatives,	103	Materia Medica, Definition of	2
Iodoform,	103	Matico,	150
Ipecac,	174	May Apple,	188
Irish Moss,	112	Menthol,	139
Iron Compounds,	44	Mercury Compounds,	57
Isinglass,	201	Methyl Salicylate,	91
Jaborandi,	181	Metric System,	3
Jalap,	187	" " Equivalents,	209
Jamaica Pepper,	133	Mezereum,	151
James's Powder,	35	Microscopy, Definition of	2
Jesuit Drops,	195	Mitigated Caustic,	71
Kamala,	198	Monkshood,	161
Kermes's Mineral,	36	Monsell's Solution,	51
Kino,	198	Moonseed, Canadian,	175
Kousso,	186	Morphine,	179
Labarraque's Solution,	42	Mountain Balm,	186
Lactose,	110	Mucilaginous Substances,	110
Lactucarium,	150	Musk,	201
Ladies' Slipper,	145	Mustard, Black,	159
Lanoline,	114	" White,	159
Lard,	114	Myrrh,	191
Laudanum,	178	**Napthalin,**	104
Lead Compounds,	53	Napthol,	104
" Water,	54	Neutral Mixture,	67
Lemon Juice,	198	" Principles,	140
" Peel,	131	Nitrogen,	61
Licorice,	148	Nitroglycerin, Spirit,	100
Lignum Vitæ,	187	Normal Factor,	84
Lily of the Valley,	145	Nutgall,	197
Liniment, Compound Mustard,	136	Nutmeg,	133
Linseed,	115	" Butter,	133
Litharge,	54	Nux Vomica,	176
Lithium Compounds,	52	**Oil,** Bay	132
Liver of Sulphur,	64	" Benne	117
Lobelia,	175	" Bergamot	122
Logwood,	197	" Birch, Volatile	122
Lugoll's Solution,	44	" Bitter Almond	119

	Page		Page
Oil, Bitter Almond Artificial,	119	Pepper, Black,	193
" Black Mustard	135	" Wood,	161
" Cajuput	123	Peppermint,	132
" Castor	117	Pepsin,	93
" Cod Liver	115	" Saccharated,	93
" Cotton Seed	115	Petrolatum,	101
" Croton	117	" Derivatives,	101
" Erigeron	127	" Ether,	102
" Ethereal	99	Pharmaceutical Definitions,	5
" Juniper	129	Pharmacognosy, Definition of	2
" Lard	200	Pharmacopœia, Definition of	2
" Lavender Flowers	130	Pharmacy, Definition of	2
" Mace	133	Phenol,	89
" Mirbane	119	Phenols,	105
" Mustard, Volatile	135	Phenyl Salicylate,	92
" Neroli	121	Phenylic Alcohol,	89
" Olive	116	Phosphorated Oil,	63
" Orange Flowers	121	Phosphorus Compound,	62
" Origanum	138	Physostigmine,	181
" Rosemary	134	Picrotoxin,	152
" Sandal	135	Pill, Cathartic, Comp.	59
" Tar	133	" " Veg.	145
" Theobroma	117	Pilocarpine,	181
" Thyme	138	Pinkroot,	183
" Turpentine	137	Piperin,	152
" Vitriol	30	Pipsissewa,	143
" Wintergreen	129	Plasma,	107
" " Artificial	91	Plaster of Paris,	41
Ointment, Rose Water	114	Pleurisy Root,	142
Oleoresins,	191	Plummer's Pills,	36
Opium,	177	Podophyllum,	188
Orange Peel, Bitter	120	Poison Ivy,	155
" " Sweet	121	" Nut,	176
" Root,	173	" Oak,	155
Oxgall,	201	Poke Berry,	151
Oxygen,	62	" Root,	151
" Hydrate,	37	Pomegranate,	172
Oxygenized Water,	37	Potassium Compounds,	63
Pancreatin,	93	Potassio-ferric Tartrate,	48
Pappoose Root,	143	Powder Aromatic,	125
Paraldehyde,	101	" Effervescing Comp.	68
Pareira,	180	Precipitated Chalk,	39
Paregoric,	178	Prickley Ash,	161
" Elixir	178	Prince's Pine,	143
Pearlash,	66	Prune,	109
Pellitory,	182	Puccoon,	182
" of Spain,	182	Pulsatilla,	153
Pennyroyal,	129	Pumpkin Seed,	188

Purging Cassia,	108	Salt Petre,	69
Pyrogallol,	105	Santonin,	155
Pyroligneous Acid,	88	Sarsaparilla,	156
Pyroxylon,	107	Sassafras,	135
Quaker Bitters,	176	" Pith,	112
Quassia,	153	Savine,	134
Quebracho,	163	Scammony,	189
Queen's Delight,	183	Scullcap,	157
Queensroot,	183	Seidlitz Powders,	68
Quick Grass,	160	Senega,	158
Quicklime,	39	Senna,	158
Quicksilver,	57	Silicon Compounds,	70
Quinidine,	167	Silver Compounds,	70
Quinine,	167	" Leaf,	183
Quitch,	160	Slippery Elm,	113
Raspberry,	109	Sloe,	161
Rectified Spirit,	94	Snakeroot, Black	185
Red Phosphorus,	62	" Virginia	136
" Precipitate,	61	Sodium Compounds,	72
" Rose,	199	Soap,	116
" Saunders,	155	" Green	115
Resin,	189	" Bark	153
" Copaiba	186	Soda, Saleratus	73
" Guaiac	187	Soluble Glass,	70
" Jalap	187	Solution Arsenite Potass.	26
" Podophyllum	189	" Chlorinated Soda,	42
" Scammony,	190	" Iodide Arsenic and Mercury,	60
Resorcin,	105		
Rhatany,	198	" Peroxide Hydrogen	37
Rhubarb,	153	Spanish Fly,	200
Rochelle Salt,	67	Sparteine,	182
Roman Chamomile,	120	Spearmint,	132
Rose, Pale	133	Specific Gravity,	4
" Red	199	Spermaceti,	106
" Water	134	Spirit, Mindererus	33
Rottlera,	188	" Sea Salt	27
Rules for converting Thermometric Scales,	207	" Wine	94
		Spoonweed,	188
Sacred Bark,	189	Spurred Rye,	170
Saffron,	145	Squill,	156
Sage,	135	Squaw Root,	143
Sal Prunelle,	69	Stag Birch,	161
" Soda,	74	Star Anise,	129
" Tartar,	66	Starch,	106
Salicin,	155	Starches,	106
Saleratus,	65	Stavesacre,	183
Saline Mixture,	67	Stearoptens,	138
Salol,	92	Storax,	196

	Page		Page
Stramonium,	183	Turlington's Balsam,	195
Strengthening Plaster,	49	Turpentine,	194
Strontium Compounds,	77	" Canada	194
Stropanthus,	159	Turpeth Mineral,	61
Strychnine,	176	**Uva Ursi,**	160
Sucrose,	109	**Valerian,**	138
Suet,	118	Vallet's Mass,	45
Sugar,	109	Vanilla,	160
" of Lead,	53	Veratrine,	184
" " Milk,	110	Vienna Caustic,	64
Sugars,	108	" Paste,	64
Sulphur Compounds,	78	Virginia Snakeroot,	136
Sumach,	199	Vitriolated Tartar,	70
Sumbul,	194	Volatile Liniment,	32
Suterberry,	161	" Oils,	118
Sweet Almond,	113	" Salt,	33
" Flag,	123	Volumetric Analysis,	81
" Spirit of Nitre,	99	" Solutions,	84
Sydenham's Laudanum,	179	**Wade's Balsam,**	195
Syrup Hydriodic Acid,	44	Wahoo,	147
Table of Atomic Weights,	86	Water,	43
" " Incompatibles,	202	Waxes,	106
" " Contractions and abbreviations used in Prescription Writing,	205	Weight and Weighing, Defi. of	2
		Weights and Measures,	2
Tamarind,	200	White Lead,	54
Tannin,	92	" Oak Bark,	199
Tansy,	136	" Precipitate,	58
Tar,	193	" Vitriol,	81
Tar Camphor,	104	" Wax,	106
Tartar Emetic,	34	Wild Cherry,	152
Tartrated Soda,	67	Wine, Red and White,	96
Tea Ash,	161	Wintergreen,	143
Terebene,	137	Witchhazel,	198
Terpin Hydrate,	137	Wormseed, American	125
Terra Japonica,	196	" European	155
Tetterwort,	165, 182	" Levant	155
Thermometers,	5	Wormwood,	118
Therapeutical Terms, Defi. of	10	**Yellow Dock,**	155
Therapeutics, Defi. of	2	" Jasmine,	171
Thoroughwort,	147	" Parilla,	175
Thornapple,	183	" Puccoon,	173
Thyme,	105	" Root,	173
Tincture, Lavender Comp.	130	" Wash,	58
Tobacco,	184	" Wax,	106
Toothache Tar,	161	Yerba Santa,	186
Toxicology, Defi. of	2	Yolk of Egg,	202
Tragacanth,	112	**Zinc Compounds,**	79
Tulley Powder,	180		

CPSIA information can be obtained
at www.ICGtesting.com
Printed in the USA
BVOW06s0300181217
503092BV00023B/2777/P